EDWARD BURTYNSKY · QUARRIES

To Raphael and Jane Bernstein, for being early believers.

For although we are accustomed to separate nature and human perception into two realms, they are, in fact, indivisible. Before it can even be a response for the senses, landscape is the work of the mind. Its scenery is built up as much from the strata of memory as from layers of rock.

Simon Schama,
Landscape and Memory

EDWARD BURTYNSKY · QUARRIES

THE QUARRY PHOTOGRAPHS OF EDWARD BURTYNSKY

WITH ESSAYS BY
MICHAEL MITCHELL

STEIDL

CONTENTS

ARTIST STATEMENT

The concept of the landscape as architecture has become, for me, an act of imagination. I remember looking at buildings made of stone, and thinking, there has to be an interesting landscape somewhere out there because these stones had to have been taken out of the quarry one block at a time. I had never seen a dimensional quarry, but I envisioned an inverted cubed architecture on the side of a hill. I went in search of it, and when I had it on my ground glass, I knew that I had arrived. I had found an organic architecture created by our pursuit of raw materials. Open-pit mines, funnelling down, were to me like inverted pyramids. Photographing dimensional stone quarries was a deliberate act of going out to try to find something in the world that would match the kinds of forms I held in my imagination but had never seen in real life — the idea of inverted skyscrapers.

We are surrounded by all kinds of consumer goods, and yet we are profoundly detached from the sources of those things. Our lifestyles are made possible by industries all around the world, but we take them for granted, as background to our existence. I feel that by showing those places that are normally outside our experience, but very much a part of our everyday lives, I can add to our understanding of who we are and what we are doing. Ultimately, what I'm looking for are interesting places and moments to embody my poetic narrative of the transfigured landscape, the industrial supply line and what that means in our life.

There are many things that the rock face can tell us about ourselves. In dimensional quarries we have to draw the substance out in a certain pattern, with a certain order. The rock face is imprinted with that order, with our methodologies, our desire, our need. Often my approach, the compression of space through light and optics, yields an ambiguity of scale. Because people are always trying to put a human scale on things, we need to put our human perspective into these images, and our presence is dwarfed by the spaces we've created. It's an interesting metaphor for how technology seems larger than life, larger than our own lives.

Based on interview extracts from:
Residual Landscapes: Studies of Industrial transfiguration,
Interview by Michael Torosian, Lumiere Press, Toronto 2001

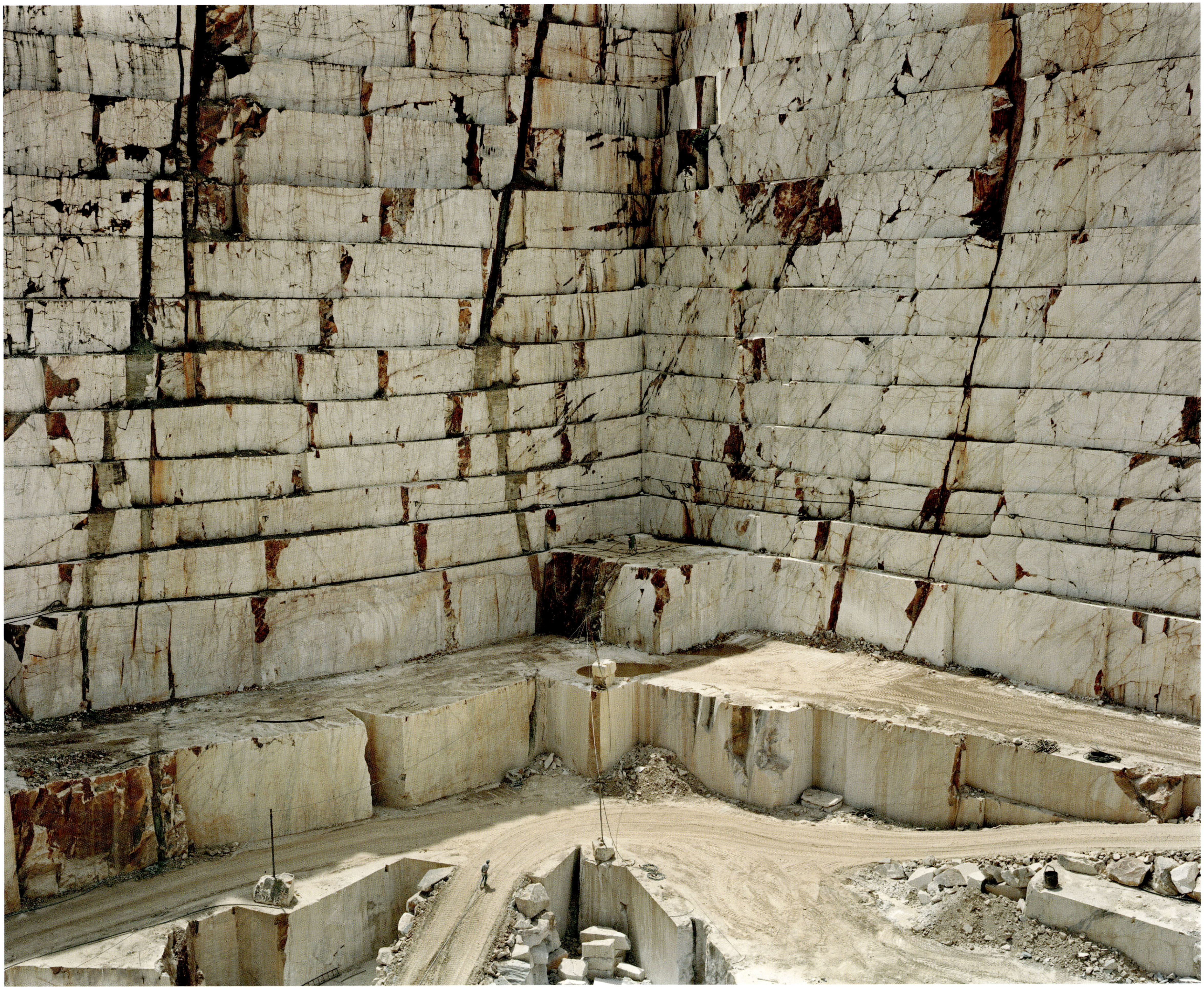

MORE URGENT THAN BEAUTY

by Michael Mitchell

I

The first people walked the shimmering plains of Africa armed with no more than digging sticks. Their only marks on the land were footsteps and shallow scratches in topsoil where they dug for roots and tubers. It was but a small step to grabbing a stone that fit the hand and pulping a fresh root, cracking a nut or crushing a shell from the sea. However, the next, attacking the rock of the land itself for material to make scrapers and projectile points, was not only much larger but also a step that threw a long shadow toward the great human alterations of the planet many thousands of years into the future.

Human beings have been in the landscape-alteration business for millennia — deforesting the Mediterranean Basin, enlarging the Sahara and creating new deserts while flooding old ones to grow more food. Over time we made an ideology and virtue out of nature-bashing and backed it up with supernatural authority. While the peoples of cultures in the East were pussyfooting through their paradise, we tough guys of the West got down to some serious land wasting, tree cutting and animal slaughter.

The Greeks laid the foundation for us. According to Socrates, the gods had provided everything for the benefit of man. Aristotle, in his Politics, declared that nature had made all animals for our sake. Classical thought generally saw humans as the bringers of order to nature. The garden was superior to the forest or open field. The sooner we got it all tamed and tidied, the better.

The Greeks' near neighbours, the Old Testament Jews, pretty much saw the whole natural world as one big opportunity. Their God had worked six days furiously building, shaping, creating, so that His people could have an entire planet to exploit.

> *"Be fruitful and multiply, and fill the earth and subdue it; and have dominion over the fish of the sea and over the birds of the air and over every living thing that moves upon the earth... Be fruitful and multiply and replenish the earth and subdue it."*
>
> *"I give you everything.... the fear of you and the dread of you shall be upon every beast of the earth, and every fowl of the air, and upon all that moveth upon the earth, and upon all the fishes of the sea: into your hand they are delivered."*

Yaweh set up the biggest resource transfer in all of history. He might be in charge of heaven, but on the earth it was open season for man.

All this sounded pretty good to the early Christians. They were definitely not part of nature. They stood aloof from it, looking down on a profane natural world that was there for the taking. Medieval Christian thinkers like Saint Thomas Aquinas refined the classical view by arguing that there was a hierarchy of being set up by God with humans at the peak of the pyramid ruling over irrational animals and domesticating them for their own pleasure and purpose. Calvin and the Protestants adopted and promoted this highly anthropocentric view of the universe. It became fundamental to European culture.

Dissenting voices, such as Saint Francis of Assisi, were too few to slow the juggernaut. Descartes's scientific method — measurement, analysis — encouraged a mechanistic approach to nature. After Newton we thought we understood everything. We seemed to get smarter; we seemed to get rich. Science was restoring the dominion over nature that mankind had lost when we got the boot from Eden. Nature was the enemy, to be exploited and defeated.

> *"[Nature's] powers are often towards man in the position of enemies, from whom he must wrest, by force and ingenuity, what little he can for his own use."*
>
> John Stuart Mill

> *"[We should] combine with the rest of the human community and [take] up the attack on nature, thus forcing it to obey human will, under the guise of science."*
>
> Sigmund Freud

The apparent returns have been impressive. Billions are being fed and we have millions of products, but as we are beginning to understand, we also face some very serious problems. We have begun to write books about them, and make films, and make art. Some photographers, Edward Burtynsky most notably among them, have devoted years to taking photographs of the huge mess we are creating.

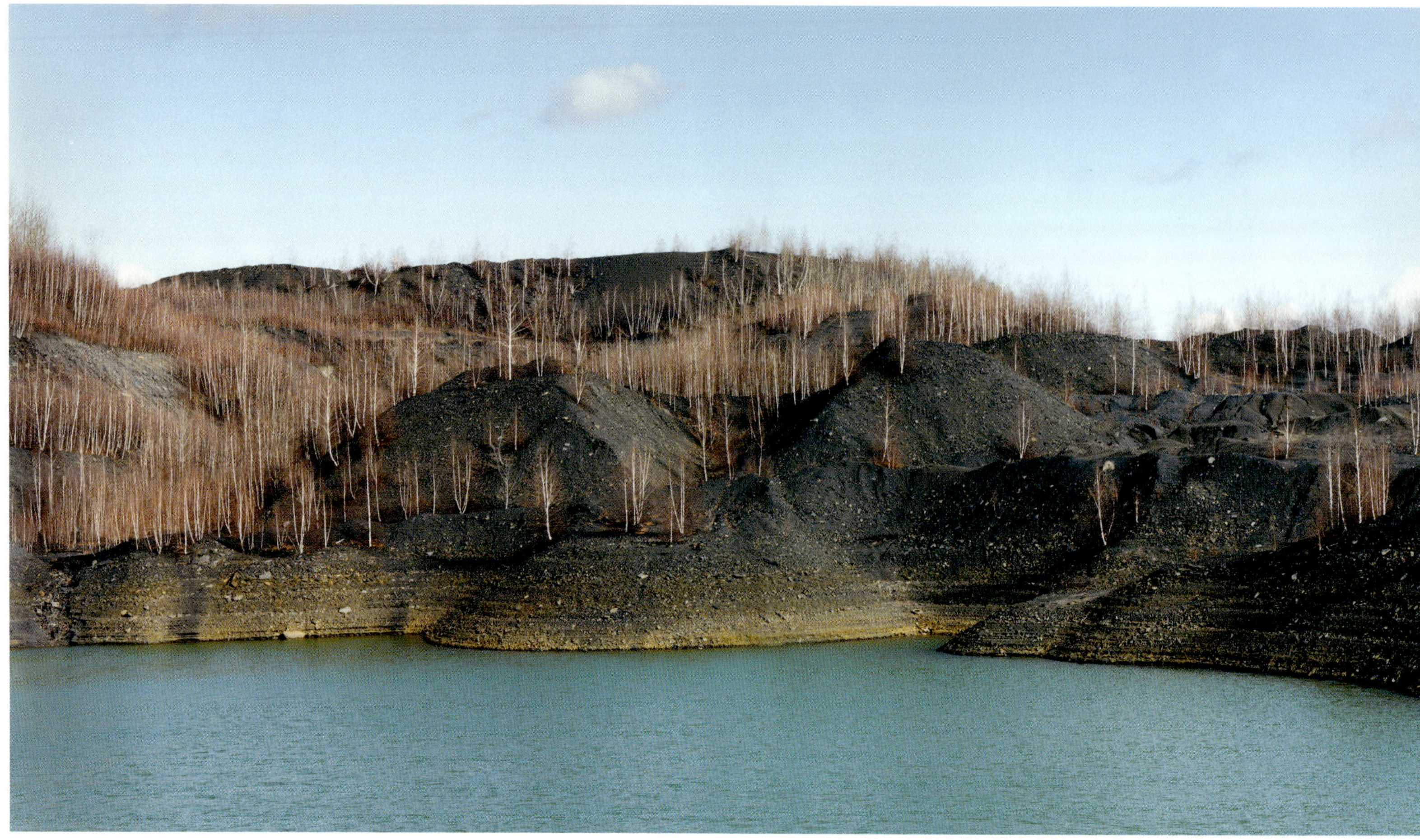

Frackville, Pennsylvania, 1981

II

One of the great rewards of a commitment to photographic image making is that it can give a serious practitioner permission to be curious about everything and everywhere. While some are energized by devoting a few months or years to a subject before moving on to a new one, others find their life's work by early or mid-career and then stay with it for decades. There are war photographers, fashion photographers, portraitists and people who only photograph birds or nudes. They often can't say exactly why they made their choice, but at some point there is a kind of epiphany, a moment when their subject presents itself to them, highlighted, as if a giant finger has pointed down from the sky and set up lights. Some subject resonates for them. It may take them years to figure out why. For Edward Burtynsky, the finger pointed twice. Like many students, he was impressed by the large-format landscape photography tradition practiced by Edward Weston. He began working to find his own voice in it. The results were competent but indifferent. Then his teacher at Ryerson Polytechnic (now Ryerson University) in Toronto told him to look for evidence of the hand of man. A small light came on, and Ed began documenting the traces of the early Welland Canals that cut through the Niagara Peninsula landscape of his native St. Catharines, Ontario.

However, it was a few years later that the really big light came on.

While travelling through Pennsylvania looking for pristine landscapes to photograph, he made a wrong turn in a place called Frackville and got out of his car. The landscape consisted of low stony hills dotted with young birches. It took him a few moments to realize that everything he was looking at was man-made. The hills were entirely composed of rock waste from coal mining. It was spooky but compelling. Here was an idea for a lifetime of work. What made him recognize it?

When Burtynsky was a seven-year-old schoolboy in St. Catharines, he joined a family tour of a nearby General Motors forging plant. It was mesmerizing and it was terrifying. Huge stamping presses and forges thundered in the grimy light. Glowing molten metal was being poured to cast engine blocks. Men moved slowly in silver heat suits. It seemed like a hell. For Ed, it was a visit to Blake's "dark, satanic mills."

It was also where his father spent much of his working life. And it would be the place that probably killed him. After years of working at the plant with tanks of PCBs, the employees began succumbing one by one to various aggressive cancers, Ed's father among them. He was only forty-six. The automaker's response was to later offer summer jobs doing toxic cleanup to some of the sons of men who'd gotten ill. By then it was clearly understood how dangerous these compounds were and how much harm any accidental spills during remediation could do to the environment. The emphatic safety lectures to the young cleanup workers made a deep impression on Burtynsky. The drastic environmental consequences of certain industrial activities became very real. He had first-hand personal knowledge of the human costs; he'd lost his father. Years later, encountering the landscape at Frackville, a part of him recognized on a deep and emotional level what he was seeing.

Now that he had his subject in broad outline, he needed the right tool for the work and some model of how to organize his pictures visually. His father had already introduced Ed to his medium. A few years earlier he had purchased a darkroom setup from a neighbour, but as a Ukrainian immigrant with imperfect English he had trouble understanding the manuals and asked his eleven-year-old son to help decipher them. A four-by-eight sheet of plywood atop a couple of forty-five gallon drums became their basement darkroom. In no time, Ed was hooked on the medium of photography.

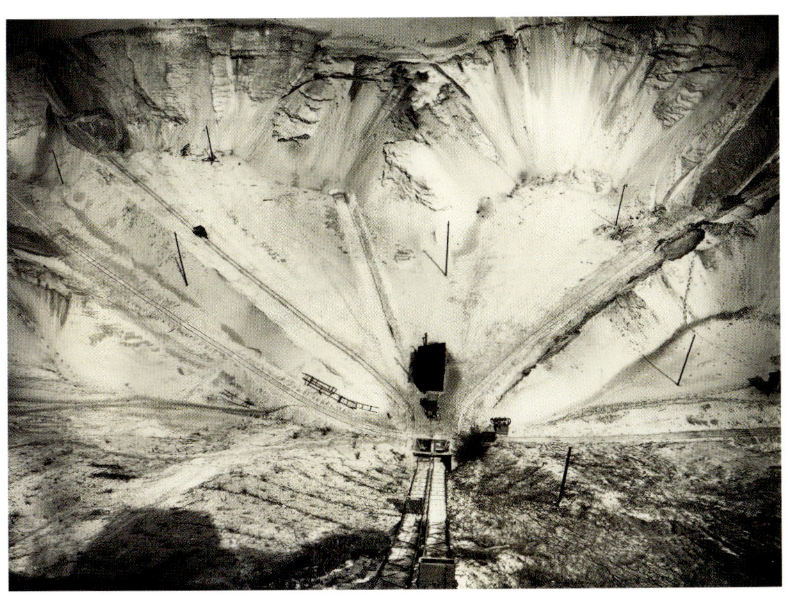

August Sander, Untitled [Quarry], ca. 1925 – 1935 (see endnote)

About a year after Frackville, at the National Gallery in Ottawa, a stunning 1932 photograph of a quartz quarry by the German photographer August Sander suggested to him how his own pictures could be structured. The Sander photograph was a forceful demonstration of the power of ambiguity; it revealed how a great gash in the landscape could be beautiful and how a photographer could both seduce the viewer and encourage reflection on the environmental consequences of our activities. Now all Burtynsky needed was the right tool for the job. His longtime friend and fellow photographer Marcus Schubert was to provide that piece of the puzzle.

III

Many of the most exciting visual discoveries in 20th-century photography were the result of cameras that used 35mm film. These miniature cameras were cheap and fun to use, and they allowed photographers to work quickly and experiment. After such masters as Kertész and Cartier-Bresson showed the amazing pictures that could be made with them, these little cameras dominated photography for most of the century. However, before George Eastman popularized roll film in the late 1880s, photographers used a different tool. The so-called view camera, on its tripod, with its large lens at the front, ground glass at the back and bellows between, was a tool of some antiquity. Long before the invention of photography, since at least the time of da Vinci, it had been used in one form or another as an artist's drawing aid. By tracing the image the lens projected on the ground glass, the artist could produce detailed drawings with accurate vanishing-point perspective. Think of Canaletto's great views of Venice and London.

These devices weren't just a cheat, they were also a great teacher that showed generations of artists how to see and how to organize and create the illusion of deep spaces on a flat canvas. They also highlighted detail and emphasized how light defined and modelled form. What the early 19th-century photographic experiments of Niepce, Daguerre and Talbot did was replace the manual tracing on paper of the projected scene with a light-sensitive plate that recorded it automatically. By the late 1800s, the negatives produced with these cameras were recording landscapes with astonishing detail and fidelity. By comparison, the ubiquitous 35mm camera of the following century merely produced a quick sketch. It painted the landscape with only the broadest of strokes.

There is a kind of symphonic theatre to view-camera photography. The photographer lugs his heavy gear to the appointed spot. He'd better have thought about it seriously because relocating is sure to be a chore. Setting up the clunky tripod and mounting the big camera is in itself a statement and a commitment to seriousness of purpose. If all photography is at heart a kind of elaborate pointing, a way of singling out that says, "Look this way at that! Isn't it interesting, beautiful, astonishing?" then using a large-format view camera is like pointing in bronze, or marble or oils. It commands us to pay attention.

The experience of the photographer using a view camera can be very intense. The artist pulls the dark cloth over his head and clicks open the lens. Suddenly the world comes swimming up onto the ground glass. And it's a world made strange by the fact that it's upside down, reversed right to left, rendered on a flat surface and glowing like an illuminated aquarium in a dark room. It's a theater of the landscape. Using these instruments can be very seductive and addictive, particularly when their unique qualities are a perfect match for your intentions and sensibility. By comparison, Ed Burtynsky's experience of 35mm photography, using his father's Minolta range finder, was more like looking at the world through a keyhole. So when Marcus Schubert lent him his 8x10 Deardorff field camera, he gave Ed his equipment epiphany. Here was the perfect tool for the job: an 80 square inch negative! Burtynsky now had his instrument.

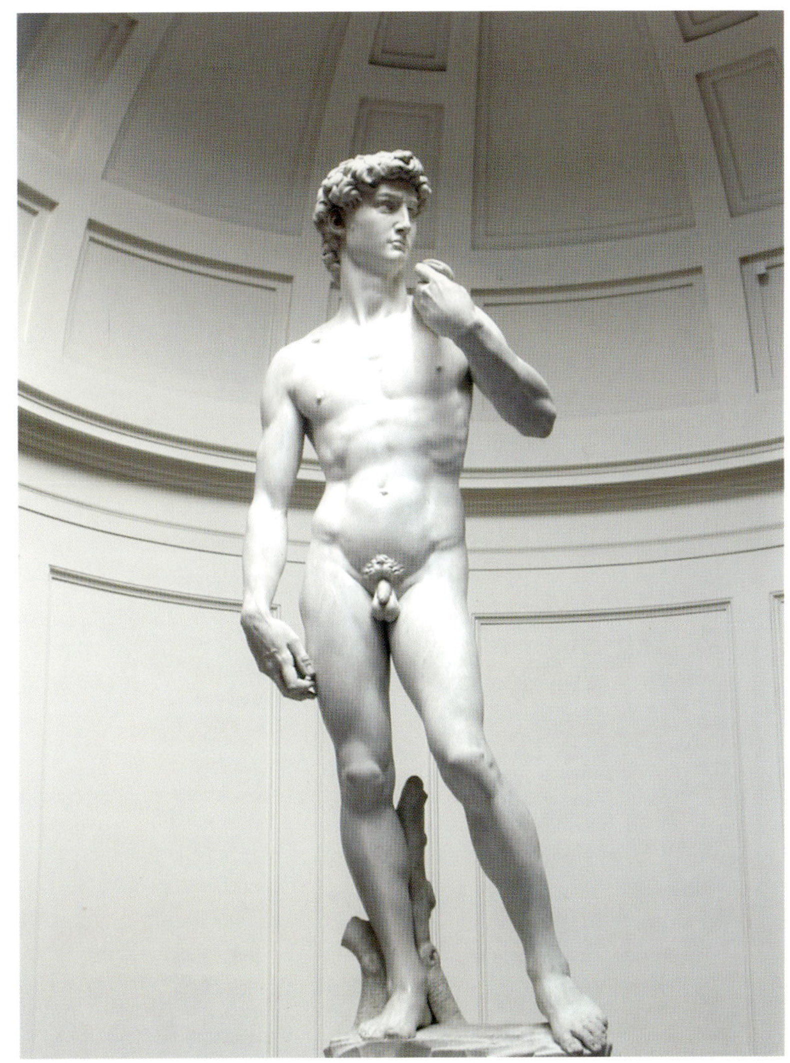

Michelangelo's David, Galleria dell'Accademia, Florence, Italy

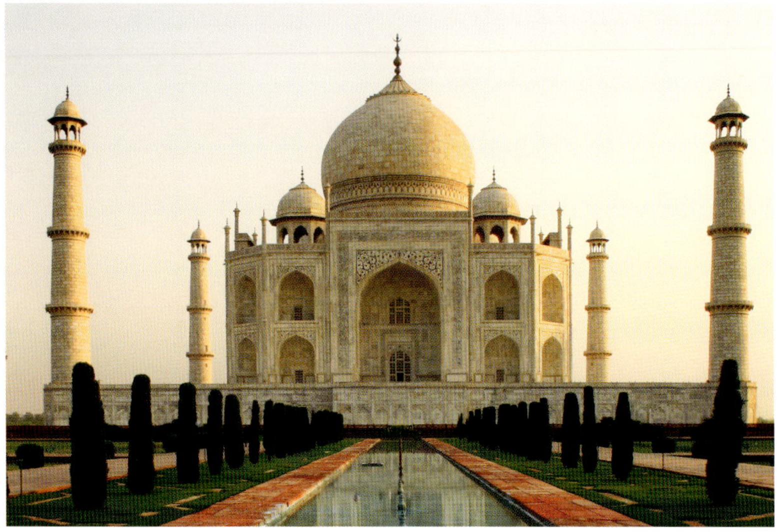

Taj Mahal, Agra, India (see endnote)

IV

Down in the White River country of southern Indiana there's a place near Oolitic where you can walk across a field and stare down into a hole that the Empire State Building came from. This abandoned quarry has steep, stepped sides framing a big puddle of murky water. All around the rim grow the kinds of scrubby softwoods that like to sneak in after men have made a big mess of the land. Nearby is another big hole because the U.S. Department of Defense decided it needed to build the Pentagon for its generals to plan wars out of the rain. When you think about it, there's a big hole somewhere for every stone building on the planet. There are holes on account of the Pyramids, the Parthenon, the Pont Neuf and the Houses of Parliament. And there are holes because we like marble counter tops, cobblestone streets and granite gravestones. Missing pieces of the Apuan Alps allowed Michelangelo to carve il gigante, the six-metre-high symbol of Republican Florence that we know as David, as well as his Pietà and his Moses.

The overwhelming majority of the quarries documented by Ed Burtynsky are dimension stone quarries as opposed to those committed to crushing rock for aggregates. Dimension stone is natural material that has been cut and finished to predetermined sizes. It is used for building, interior finishing, pavers and monuments. By and large, this means one of two types of stone: igneous rocks or those of metamorphic origin. Granites are very hard crystalline rocks composed mainly of quartz, feldspar and mica. They are igneous because they were born of great heat, melting and pressure. Burtynsky shows us granite quarries in Vermont and eastern China. By contrast, his quarries in Italy, India, Portugal and Spain all produce marble. These metamorphic stones are sedimentary calcium carbonate that has been crystallized by heat and pressure. Marbles are softer than granites and considerably more porous. Buildings in harsh northern climates that have been clad in marble tend to develop problems with time. Marbles exposed to freeze-and-thaw cycles can warp, crack and fall down. There are tall, marble-clad buildings in Chicago and Toronto that are in trouble. But in more forgiving environments, marble can add much beauty to a building. Marbles from Makrana, in India, are so translucent that you can see light and shadow through slabs two inches thick. That's why the Taj Mahal glows. Marbles from Carrara share some of this property. Sculptures carved from statuario, the finest and whitest grade of Carrara marble, tend to dazzle and shimmer in the light. And marble exteriors in temperate or tropical environments can last a very long time.

Compared to the impact on the landscape of other mines and many industrial processes, quarries may seem fairly benign. Quarrying leaves no acids or heavy metals, only rock dust, stone shards and large holes. Burtynsky's Vermont photographs show us that nature quickly moves in on abandoned quarries to plant trees and shrubs and install reflecting pools. One of the surprises provided by re-photographic survey projects in the American West was the revelation that the land often looked better in the late 20th century than when it was first photographed in the 19th. We like to think of the great large-plate landscape photographers like Muybridge, Watkins and Jackson as bearing photographic witness to pristine landscapes west of the Mississippi, but those men travelled west largely by train, and the railways were built because people were already busy digging up the territory. The trains had come to move labour and supplies in and ship the spoils out. When, in the 1970s, some American photographers carefully set up their tripods on the same spots as the pioneers had a century earlier, they discovered that the wounds of the 19th century had often weathered into invisibility and that there were frequently many more trees. The land was a patient patient. With time it seemed to recover.

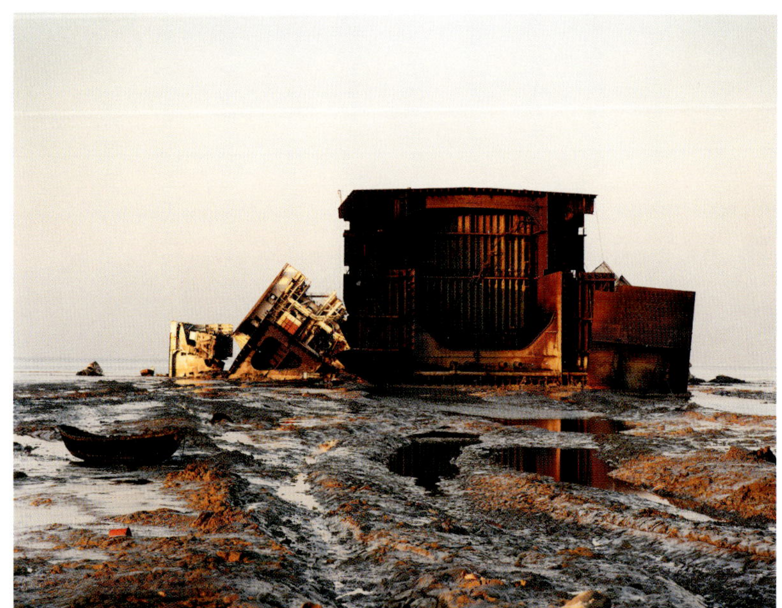

Shipbreaking #1, Chittagong, Bangladesh, 2000

The first exploration photographers of the American West gave Burtynsky ideas about how to deal with certain compositional challenges that landscape photography always presents. How does the photographer visually resolve the foregrounds in landscape pictures? In a pinch, a painter can always drop in a foreground tree to frame a distant view. The photographer, especially one in the Wild West, was more often stuck with a barren foreground of small rocks and dry, scuffed dirt. Since landscapes are photographed focussed at infinity or close to it, those boring foregrounds were often out of focus as well. They added nothing to the sublimity of the great sweep of hills and mountains in the middle and far distance. They became a visual handicap. The 19th-century masters dealt with this problem by setting up on small banks and hillocks; if none was available, they used a wagon. If there was no wagon, they built a platform. The foreground then retreated into the middle distance. Problem solved.

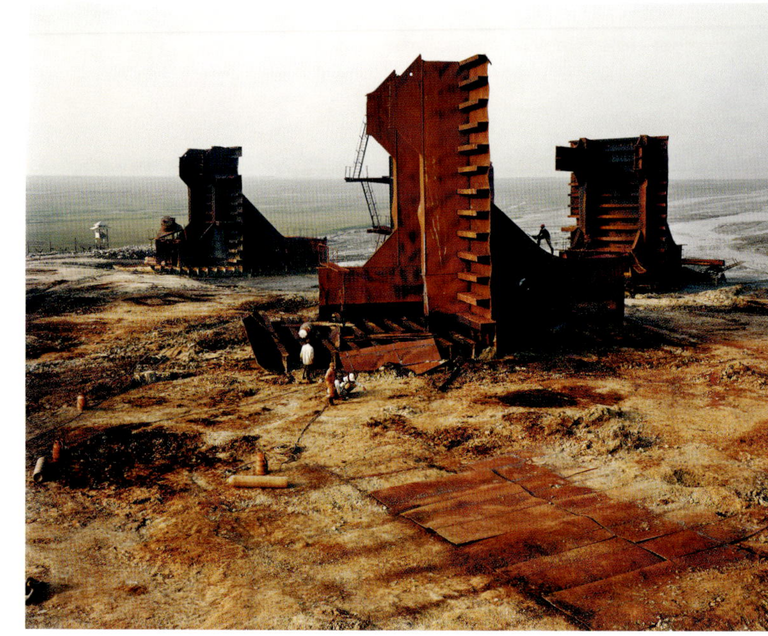

Shipbreaking #27, Chittagong, Bangladesh, 2001

One can see Burtynsky encountering this issue in some of his early pictures. At least one of the early Chittagong shipbreaking images that he exhibited at his first Toronto gallery had great clumps of out-of-focus muck in the foreground, but all the Chittagong pictures later reproduced in his book Manufactured Landscapes are taken from somewhat elevated vantage points. They take us into the scene without foreground distraction.

Quarries, however, have inherent spatial qualities that often resolve this issue for the photographer. Photographing the great sweep of an open pit from the opposing rim eliminates the foreground. So does aiming straight down into it. When neither of these situations was possible, Burtynsky unapologetically availed himself of the tools and tricks used by industrial photographers — scissor lifts, hydraulic buckets, anything that would go up, to get the perfect vantage point for his field camera.

Once he began stepping off the North American continent to pursue his themes, Burtynsky had to develop a very systematic approach to each project. He could no longer be the solitary artist driving the countryside hoping to find landscapes that talked to him. He had to do research, hire scouts, translators and negotiators. He was now running expeditions in which the time spent actually taking pictures was overwhelmed by the time devoted to advance planning. In Toronto, his home base, he found it necessary to establish his own commercial colour lab so that he could control the quality of the enlargements he released. His personal photography, after his first offshore session at Carrara, began to take on an industrial scale of its own.

V

In general, Burtynsky has used very traditional pictorial means to make his quarry pictures. The deep depth of field is from the 19th-century book of landscape photography and such 20th-century disciples as Edward Weston and Ansel Adams. So are the notions of how to organize elements of the subject within the frame. There are no experiments with tilted horizons, unusual lighting, radical uses of focus or image manipulation. Instead we have serene, classical technique that everyone understands. It's as if the photographer wished to make as transparent a window as possible, so that the viewer can be taken straight into the scene and immersed in its unsettling implications. Ed unashamedly uses the raking romantic light of dawn and dusk to enhance his industrial landscapes. And he schedules his excursions for the transitional seasons — spring and fall — when the light is most seductive.

Burtynsky began his quarry series by looking at something he has called the "residual thing" — that is, what remains after man is finished with something and walks away, leaving to nature the job of slowly pulling the abandoned object back into the ground. Photographing these melancholic sites has long been meat and potatoes for painters and photographers. They try to record or fabricate a kind of romance for these places. This is not all that hard to do. However, to advance into living sites, ones that are still being worked, and make art from them, is considerably more difficult. The resulting pictures can easily look like mere reportage or annual report photography. Burtynsky has taken on that challenge, and by and large he succeeds.

The photographer has remarked that making great pictures is not a matter of numbers — if you make enough exposures, one will eventually be good — but of hard-won professional practice and artistic commitment. The large-format photographer-artist works slowly and deliberately, bringing the experience of years to every exposure. Photography is basically an editing medium. Selecting a quarry section to frame inside the ground glass during the light of early morning involves two edits—the exclusion of all other parts of the scene and all the other times of the day. After these initial spatial and temporal edits have been made, there are still tough ones to face back at the studio and in the darkroom. The photographer can edit tones and light by burning and dodging during printing or by manipulating files. It can take a very long time to get a print right, and even then, selection for formal coherence or contribution to a series or a theme can result in a handsome, well-made print being edited out. Behind the big seductive print on the gallery wall lies a ranked sequence of very tough choices. Classical photographers can't over-paint, diddle with scale and dabble in colour changes. Ultimately they are reality's captives. They have to make peace with the world as it is.

Edward Burtynsky has worked very hard to make peace with his somewhat hardscrabble past. His father Peter's long journey took him from the Carpathian Mountains of Ukraine through the relentless bush of Kapuskasing in northern Ontario to the gritty world of heavy industry in a border town and an early death. Back in Ukraine, the family had owned a tile factory and made beautiful things. The young immigrant to Canada took art lessons by correspondence. All he ever really wanted to do with his life was be an artist and paint pictures. In reality, the closest he ever got was painting stage sets on the weekends for the local Ukrainian hall. However, his son got to complete the ambition and take the journey, working for a time in mines and auto plants himself in order to go to art school full-time and then take on the long apprenticeship as an image-maker until finding his true subject and his own voice.

He first found both when he began his quarry series nearly two decades ago. Since then he has made photographs of quarrying operations in a half dozen countries. His camera has also explored other kinds of mining, various recycling operations and large-scale manufacturing. Those pictures must inevitably inflect how we see his quarry images. If there is a narrative running through all these photographs, Burtynsky insists that he wants to keep it open. He doesn't want the summation of his work to be merely a lecture on environmentalism and the profligacy of consumer culture. Clearly he can find beauty in ugly things. And he celebrates that tension he generates by anesthetizing carelessness and waste.

As artists mature, they become less concerned with control of meaning and more appreciative of multiple interpretations. They learn from them and build upon them. After working in Carrara in the early 1990s, Burtynsky found himself thinking about our marks on the land from a more global perspective. His experiences in India photographing quarries and in Bangladesh recording shipbreaking shocked him into a reconsideration of human existence. He witnessed people living on very small means. They owned little and their lives were short. What should issue from the encounter between a now-privileged North American with an expensive camera and an exhausted South Asian squatting in a Makrana quarry breaking rocks by hand twelve hours a day, six days a week for seventeen years? Burtynsky is still thinking about that encounter.

While Burtynsky's essential concerns are obvious — the consequences of unbridled development and consumerism, of utilitarianism and short term economics, of the First World leaving the dirty work to the Second and Third — he is careful to try and leave the door open to wider readings. He can only hope that the friction between the romantic vision he brings to the world of extraction and waste, of consumption and disposal, generates dialogue and in some way points toward change.

We live in a wealthy and privileged time and place where many thousands more students go to art schools than was ever possible before. They learn to be professional, to be ambitious and to aspire to international careers. They read the journals and attend exhibitions as if they were trade shows whose chief utility is the opportunity to scout out the competition. The work these young, university-trained artists do is current, professional and career-strategic. It is often admirable, but much of it fails to talk to a wider, intelligent public. Its issues are not theirs.

Burtynsky manages to step over the pretty fence around the art world and touch a much larger public. A critic might say that is because the work relies on an established pictorial aesthetic that almost everyone understands. We have all internalized the paintings or sculptures of Friedrich, of Turner, even of Pollock and Chamberlain and a host of other artists that may appear to be influences. However, countless other artists have successfully absorbed these artists' works yet have not garnered the public that Ed Burtynsky has. His international audience recognizes something more.

It would seem that what they respond to is an authenticity. The virtue and strength of Burtynsky's work is that the impulse driving it comes from deep experience — a background of hard labour in workers' towns, a wilderness camper's passion for the natural world and an experience of loss close to home. This is what makes it true. His compulsion may have come from personal biography, but he has managed to use it to step up to something much larger than himself. In a time when a lot of work is handsome and seductive but ultimately little more than a skillful product, Burtynsky has managed to talk about something more urgent than beauty. He has made us witness to the consequences of our cornucopia. He shows us what happens when man, once cowed by nature, now finds himself overwhelmed by his own interventions in it.

There are thinkers in the West who dream of a future humane secularism in which our species faces absolutes like death and entropy without fear, faith or fables: a future in which we realize that all that we see around us now is all that we are going to get. We had better walk softly and take care of the world we have because there isn't another one up in the sky. Until we do, Burtynsky will have no shortage of sites to photograph.

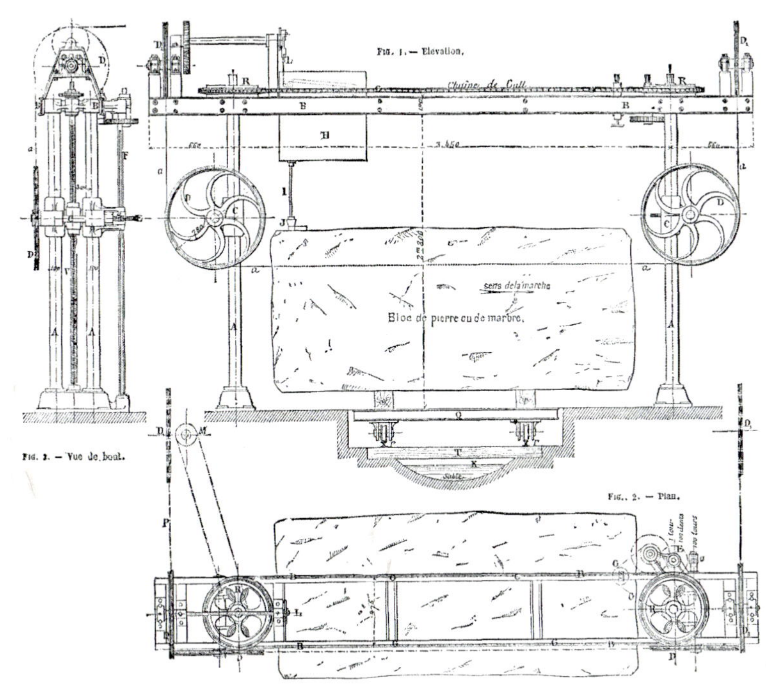

Helicoidal Wire Stone Saw, 1880s (see endnote)

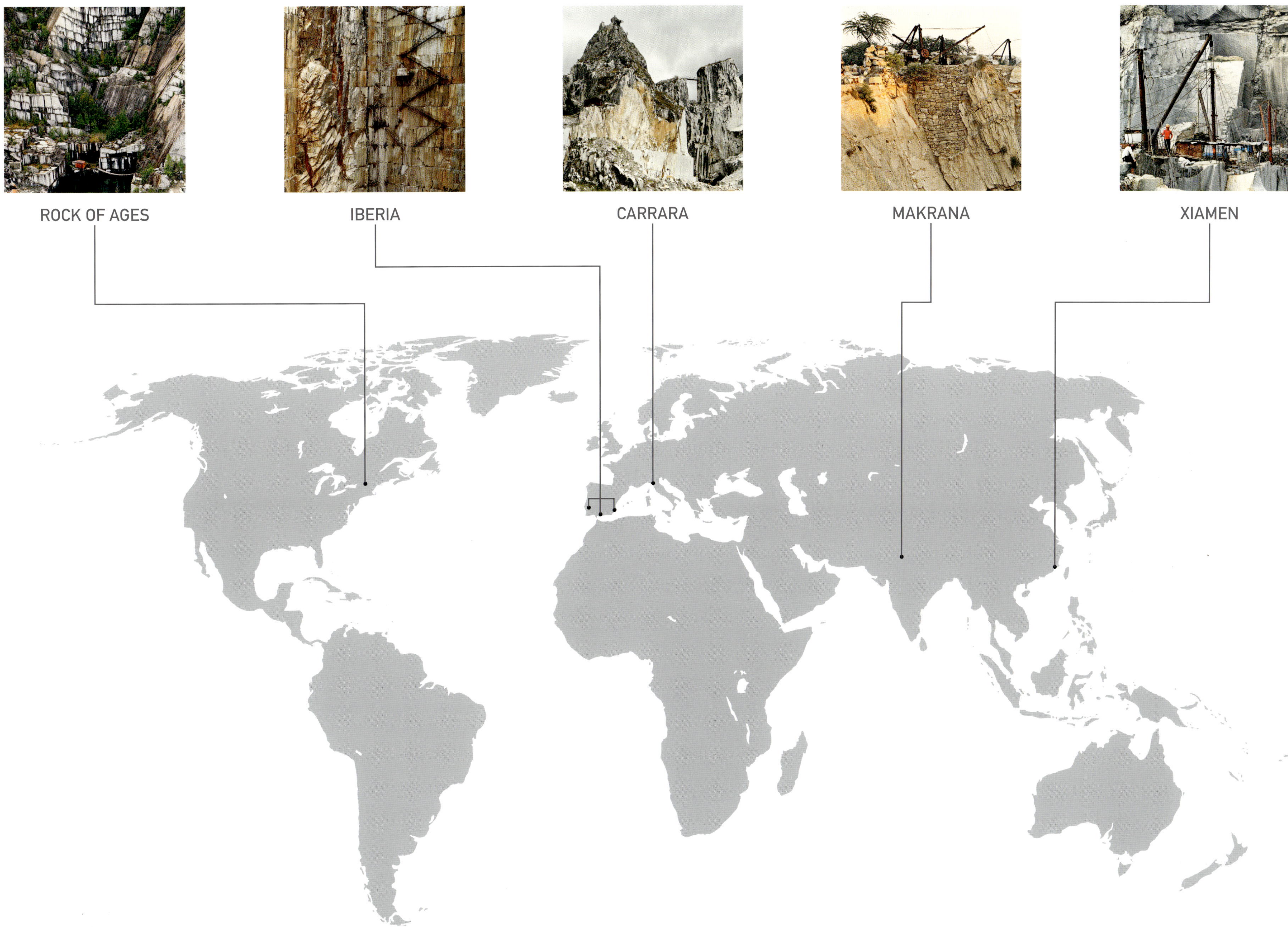
ROCK OF AGES
IBERIA
CARRARA
MAKRANA
XIAMEN

VERMONT - ROCK OF AGES

Over the past several years, business at Rock of Ages Corporation has not been good. North America's largest manufacturer of granite memorials lost $3.2 million in 2004 and more than $16 million in 2005. Part of the problem has been a tightening of credit in China's construction sector, which has depressed demand for Vermont granite in the Far East. However, the big factor is that 1933 was a low-water year for live births in the United States. With an average life expectancy of seventy-three years for Americans, there just aren't enough people dying these days to make the tombstone business profitable.

Rock of Ages is a huge enterprise. It operates ten active quarries and five manufacturing and sawing operations, principally in Vermont and Quebec. Its granite products are sold through more than eighty company-owned retail outlets.

It wasn't always so big. The town site, in the granite mountains of Vermont, was first settled in 1788 and named Williamstown, then, more awkwardly, Wildersburgh. In 1793, a legendary and perhaps fictitious fistfight between two prominent locals promised naming rights to the victor. He chose Barre, after his hometown in Massachusetts, and Barre it's been ever since.

The town sits on a one-by-three-mile deposit of gray granite in the 500-million-year-old Vermont mountains. The first documented use of Barre granite was to make millstones and road pavers in the late 18th century. Soon after the War of 1812, Barre got its first professional quarryman, Robert Parker, whose quarry is still worked today as part of the Rock of Ages operation. The town's big break came in 1832, though, when the Wheaton quarry on Cobble Hill was contracted to supply building stone for the new Vermont State House in Montpelier.

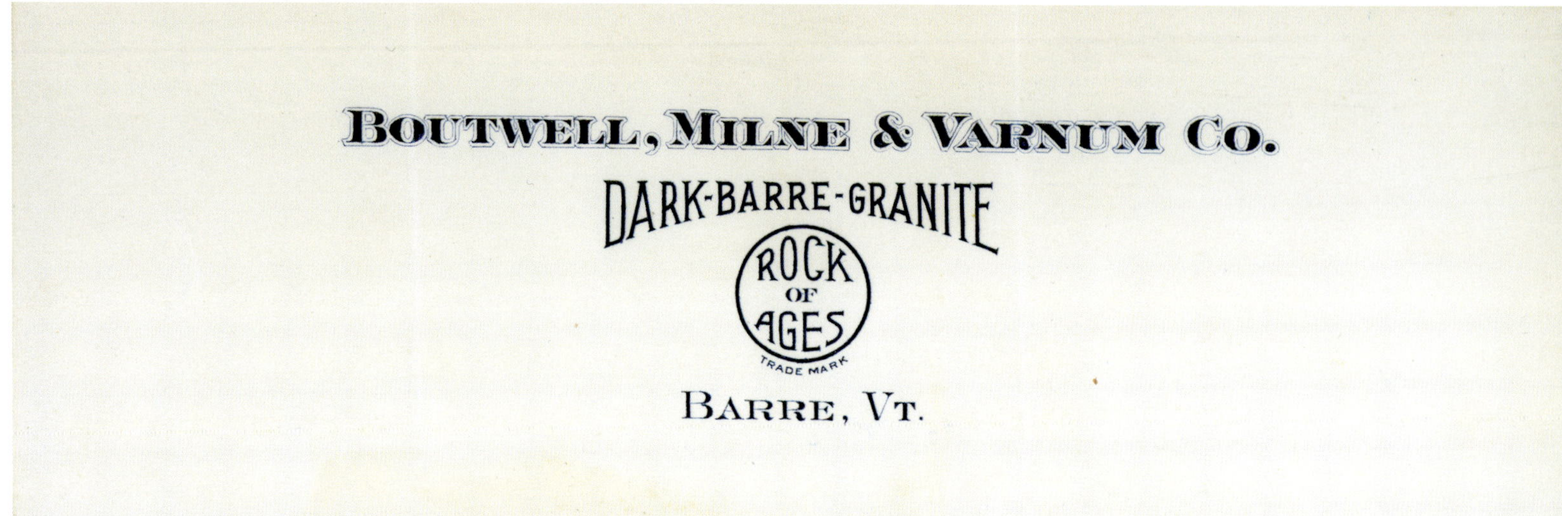

But while nobody disputed the quality of Barre's granite or the skills of its workers, for years transportation problems kept it on the sidelines of the dimension stone industry. It was high and it was remote. Up to forty draft animals were required to move a granite block. A fifty-ton block could only be transported at the rate of a mile a week. Consequently, by the time the American Civil War ended, there were still fewer than ten men working the Barre quarries.

In 1875, the Vermont Central railway finally reached Barre to the cheers of six thousand people. It would be another twelve years though, until the *Barre Quarries Spur* made the four-mile climb up the one thousand feet of Millstone Hill to the actual quarry sites. Then the Barre dimension stone business boomed (despite even the downhill, runaway crash of a granite-filled train in 1893). The town's population more than doubled each decade between 1880 and 1900. A wave of Scots quarry workers from Aberdeen was followed by Italian stone cutters and then Swedes, Spaniards, English, Irish, Greek and French-Canadian quarriers and carvers.

Business thrived and working conditions changed. After the 1885 arrival of electricity made possible the extensive use of power tools of all kinds — electrical, steam-driven and pneumatic drills, hammers saws and surfacers — the pace of work accelerated dramatically, productivity rose, and the workers began dying. There seemed to be funerals every week. The unions blamed the owners and the owners and the medical establishment blamed the workers. Everyone agreed that the quarrymen were dying of lung disease. For the labourers in the carving sheds, the source of the problem was obvious. The new power tools were filling the poorly ventilated buildings with fine silica dust, which eventually suffocated them. The owners and their doctors claimed the problems came instead from the slovenly and unsanitary conditions in Barre's North End where the workers lived — a perfect breeding ground for the newly discovered *Tubercle bacillis*.

Years of labour strife followed, both strikes and lockouts. Barre became an activist hotbed, with regular visits by the likes of Eugene Debs, Norman Thomas, Mother Jones and Emma Goldman. During a major granite workers' strike of 1922, the companies imported large numbers of French Canadians to keep the quarries and sheds open. Finally, in 1937, some $300,000 worth of ventilating equipment was installed, and silicosis, that most dreadful of occupational diseases, disappeared. The unions had been right.

Of the dozens of firms that made up the Barre granite industry in the early 1900s, Barclay Brothers, established in 1887, was one of the largest. Founded by a Scot who had originally emigrated to Quebec, Barclay Brothers was a key member of the nine firms that merged in 1930 to form Rock of Ages Corporation. (The name had been suggested by an advertising agency in Burlington hired by Milne, Boutwell and Varnum, one of the other quarry operators in the merger, which hoped to improve marketing and quality standards and establish a price structure.) Now fully one third of the Barre granite business was under

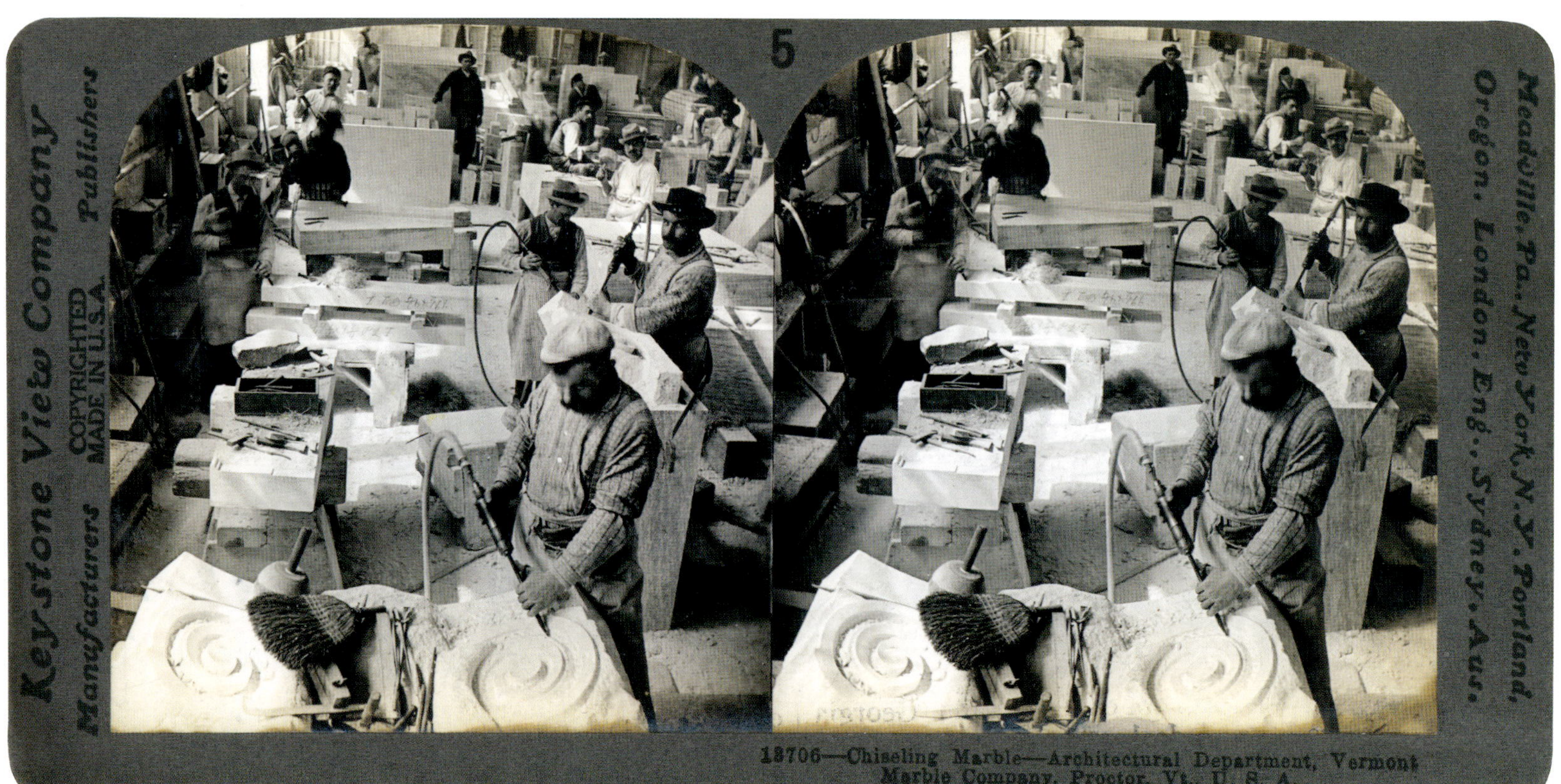

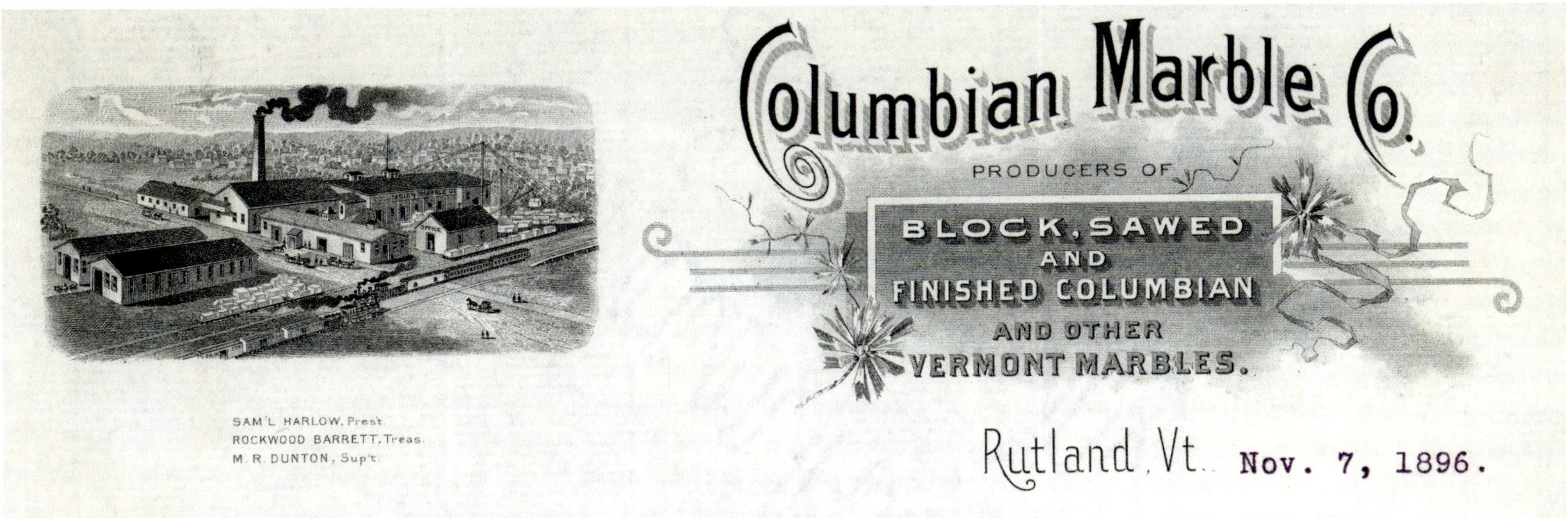

a single corporate umbrella.

Ed Burtynsky got to Barre for the first time in 1991 as a result of a photographic quest for quarries in Northern Ontario. When he expressed his disappointment with the small scale of operations in Temagami, a quarryman there described a series of spectacular quarries he'd seen in Vermont. Ed packed his gear and began the long drive southeast to Barre. It was a trip that launched his career.

With a Rock of Ages quarryman at his side, Burtynsky set up his friend Marcus Schubert's Deardorff view camera. As he prepared to expose his very first sheet of 8x10 film, his guide began teaching him how to read the towering manmade rock face (page 29). For Ed, it suggested inverted architecture: an idea about quarries that he had long dreamed of, but in the eyes of the quarryman it was an evolutionary history of extraction technology. The rows of tiny holes that the huge format recorded so well in the lower right of the image were the result of an old technology called channeling. Operators would drill a series of holes along a fault and load them with small charges of dynamite to separate the rock. Where the rock face was unstable, bolts ten to fifteen feet long had been drilled in place. These are clearly visible on the upper left of the photograph, just below the patch of sky. After the late 1950s, this traditional quarry method was replaced by high-speed cutting torches, multiple burners on a long bar that heated and popped the feldspar and quartz until it separated. In the image on page 59, one of these cutting bars is visible to the left in front of the small red shed at the center of the photograph. A puff of smoke marking a cut in progress can be seen issuing from the rock face just below a pair of orange tanks supplying the burners with bunker oil. This newer quarry has no channeling holes.

In all, Burtynsky made a half dozen Vermont trips to photograph what are thought to be the deepest quarries in the world. (Rutland's abandoned marble quarry on page 67, with its massive diagonal pillars, supports a marble face that was mined down to eight hundred feet.) These photographs tell many stories. On page 17, an enormous wooden derrick stands abandoned and marooned on a towering island of granite inside an old flooded quarry. The derrick tower and boom are four- and five-foot diameter logs of Douglas fir from British Columbia, which were shipped via the Panama Canal and carried by rail to Barre on specially constructed train cars.

A trip in early May of 1992 yielded a series of photographs including those on pages 41 and 43, where the architectonic qualities of dimension stone quarries are highlighted by a freak late spring snowstorm. While the 19th-century exploration photographers whom Burtynsky so admired had often relied on inserted human figures to provide scale, in these Barre and Rutland photographs he has learned to suggest size not by using people but by recording the steps and ladders that gave workers access to the pits.

By 1986, Rock of Ages was producing more granite each year than all of Barre's sixty-eight independent stone companies had produced in 1902. The late 1980s saw Barre companies generating $80 million in sales and employing fifteen hundred people. Barre is the only place in the United States where the granite industry is fully unionized. Just as a local firm boasted in 1905 that "No one has ever yet found the bottom of a good Barre granite quarry," it would seem that no one is ever likely to see a Barre granite worker who is not unionized. There's a long memory in this town. People have not forgotten what the unions did for dying workers in the 1920s and 30s. Nor will Burtynsky, the photographer, forget what happened in the auto plants of St. Catharines. "I see that safety is something hard fought when it's won."

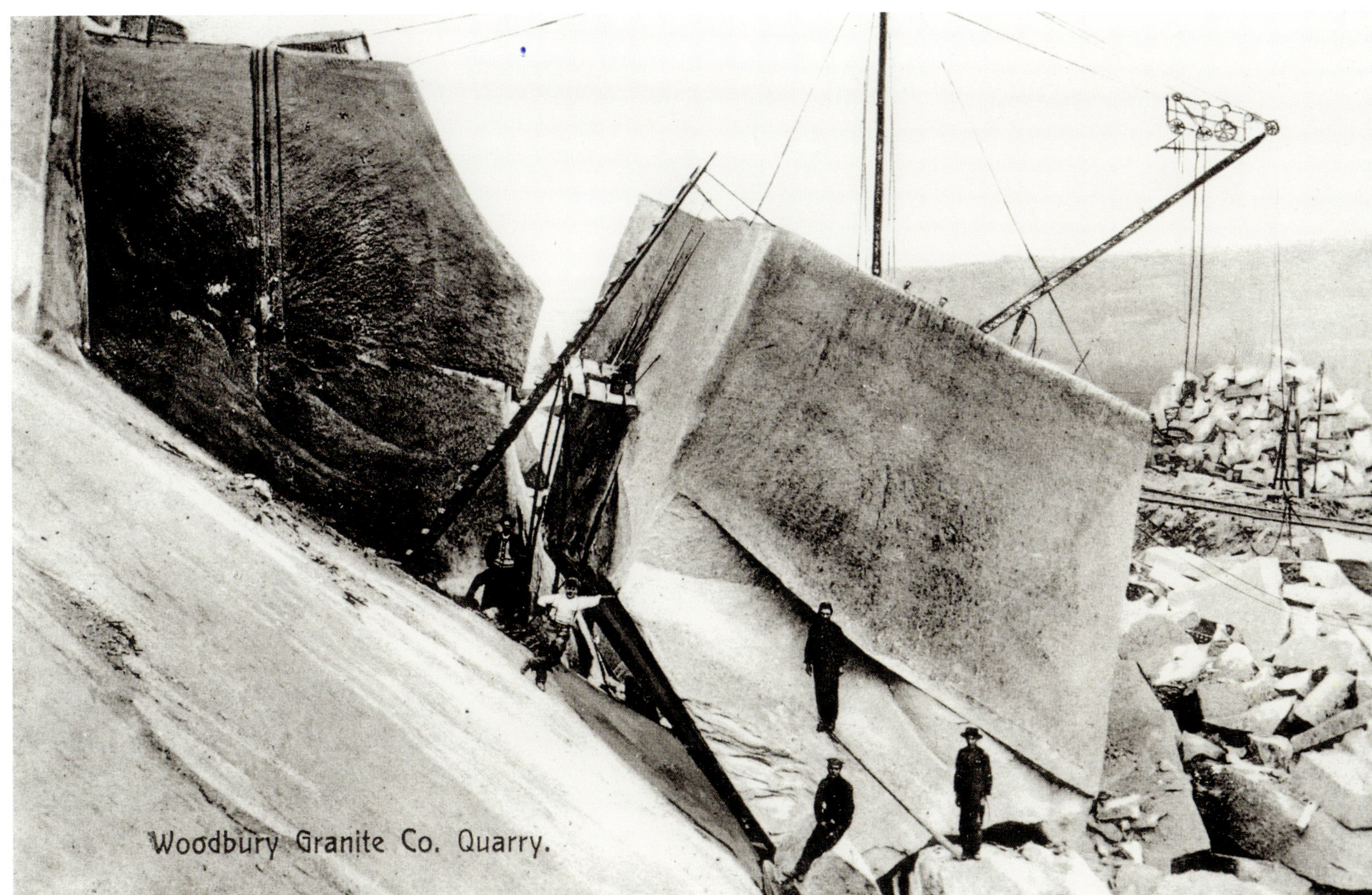

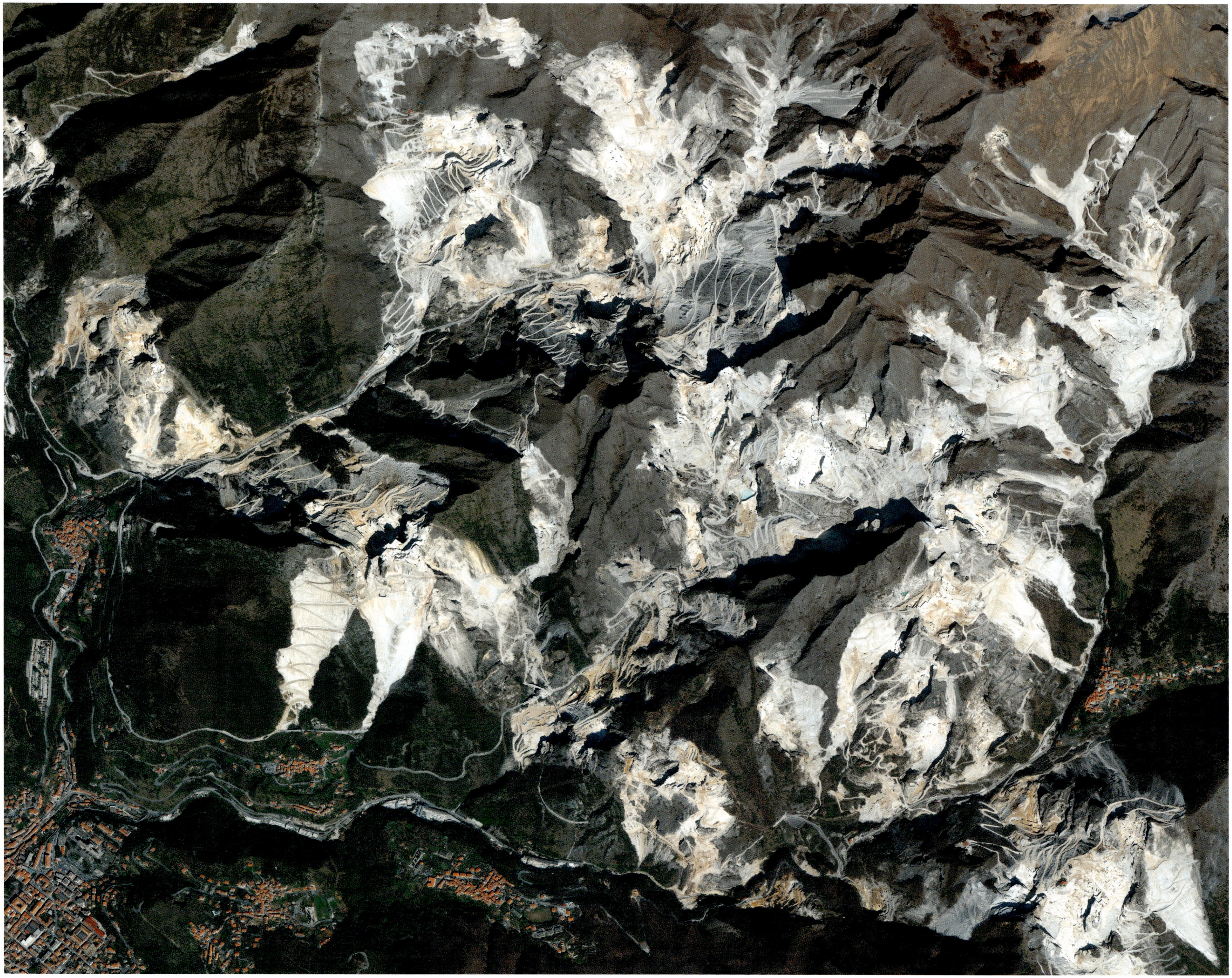

CARRARA - THREE MARBLE MOUNTAINS

At midnight on May 14, 1504, a crew of forty men smashed through the wall of the duomo workshop in Florence and rolled Michelangelo's seventeen-foot high statue of David out into the dark streets of the city. The five-block journey to the Palazzo della Signoria would take four days and must have seemed more like the aftermath of a European soccer match than the installation of one of the Western World's greatest works of art. The citizens of Florence had recently overthrown the ruling Medici to re-establish a republic. For many Florentines, the giant statue was a symbol of this new order and those citizens who still supported the ousted Medici spent much of David's four-day journey pelting the enormous marble statue with rocks.

The block of Carrara marble that became David had been sitting in the duomo yard since 1466, almost forty years. Agostino di Duccio, the first sculptor to tackle the block, had been sacked in December of that year. A decade later the commission was briefly taken on by Antonio Rossellino, who roughed out a figure but then gave up. The block had developed hairline cracks after falling off an oxcart when it was first delivered. It was fragile because, as chunks of Carrara marble go, it was garbage. Experienced carvers worked only with *statuario* — the highest grade of hard, flawless, pure white Carrara marble. The duomo block was the porous and pitted stone the quarrymen called *ordinario*. It was so mediocre that the twenty-six-year-old Michelangelo had gotten it for free.

When the Romans first began erecting significant buildings, they imported all their fine marble from the Greek islands of Paros and Naxos. It wasn't until the first century A.D. that they discovered marbles of superlative quality in their own backyard. Today, at least thirty Roman quarries are known in the mountains behind Carrara, including the great and perfectly preserved Fossacava. Although quarrying at Carrara almost ceased during the Barbarian invasions, the spread of Christianity after the fifth century, with its demand for sacred buildings, revived and greatly expanded the mining of marble in the Apuan Alps. A whole culture of specialist workers evolved: the *caesores*, who detached the blocks; the *quadratari*, who shaped the marble, and the *maquinarii*, who transported the huge blocks using wooden carts and giant ropes. Charles Dickens, who visited Carrara in the mid-19th century, described teams of up to twenty pairs of oxen dragging the blocks down the mountains to the sea. Their drivers jammed huge pointed rods of iron into the bodies and nostrils of the oxen to keep them mad with pain and moving over the rough stones of the streambed that formed the main road down the mountain. Many oxen died this way. When Dickens subsequently visited the great statue-carving workshops in Carrara, he marvelled that "these exquisite shapes, replete with grace and thought and delicate repose, should grow out of all this toil and sweat and torture!"

Over the centuries, many great artists have carved marble from Carrara. Leonardo da Vinci designed a special saw for the quarries. Michelangelo made several trips to the region to personally select blocks for such projects as the Pietà in St. Peter's and his Moses. There are sculptures carved from Carrara *statuario* by Canova, Henry Moore and Louise Bourgeois. The architects known as the *maestri comancini*, who descended on Carrara during the great church-building years, made the local marble key elements of the churches and domes of Florence, Siena, Pisa, Orvieto, Genoa and Lucca. They were continuing a tradition that reaches back to the Pantheon and Trajan's column in Rome and abroad to London's Marble Arch and beyond. The view of Toronto's financial district from Ed Burtynsky's studio is dominated by a half dozen bank towers, including architect Edward Durrell Stone's seventy-two storey First Canadian Place, a building clad entirely in Carrara marble.

Like many of us, Burtynsky went to Carrara dreaming of Michelangelo. But after a long flight and an even longer mountain drive he found the road to the master's favourite quarry chained off a dozen kilometres from the source. His view of that historic marble mountain from across the adjacent valley (page 79) records the closest he ever got. Yet frustrated plans can yield unexpected opportunities. A nearby quarry suggested as an alternative proved to be a picture-maker's windfall. The creamy, flawless marble made a perfect white ground for the machines, cables and tools of the quarryman's trade. In the diptych (pages 104 and 105), the chainsaw-like block cutters slowly make their way along short track sections on the quarry floor. It's as if they're cutting cake.

The cutting method most often employed at Carrara is by diamond cable. The hole drilled to feed the cable through the rock can be seen at the apex of the right angle cut a few metres right of the ladder (on page 99). The white machine that winds the diamond cables through the stone rides on a short track at the bottom centre left on page 95.

Today there are over one hundred active quarries around Carrara. They employ a thousand workers who extract a million tons of marble each year.

EB

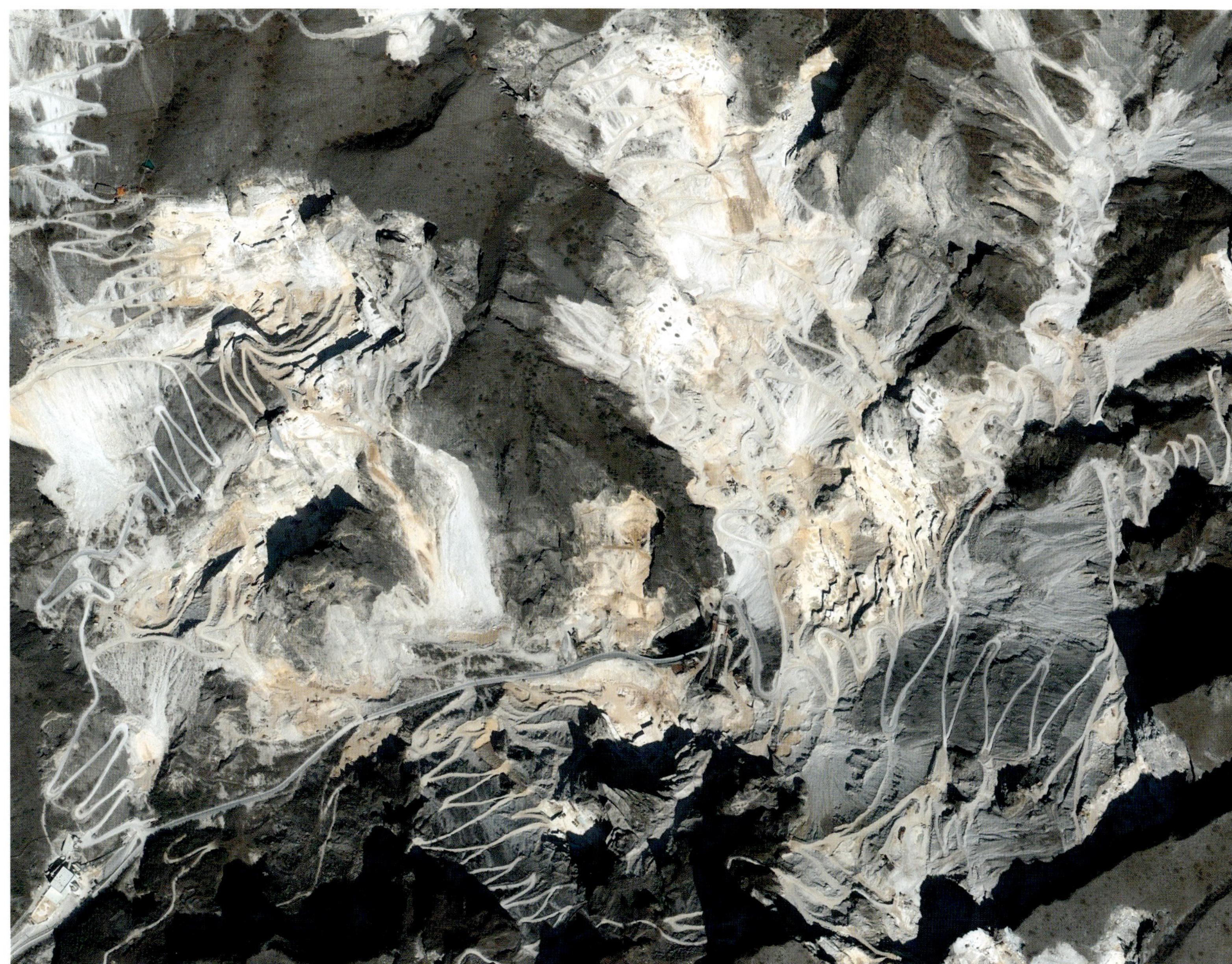

Carrara Marble Quarries, Satellite View (see endnote)

MAKRANA - DYING FOR BEAUTY

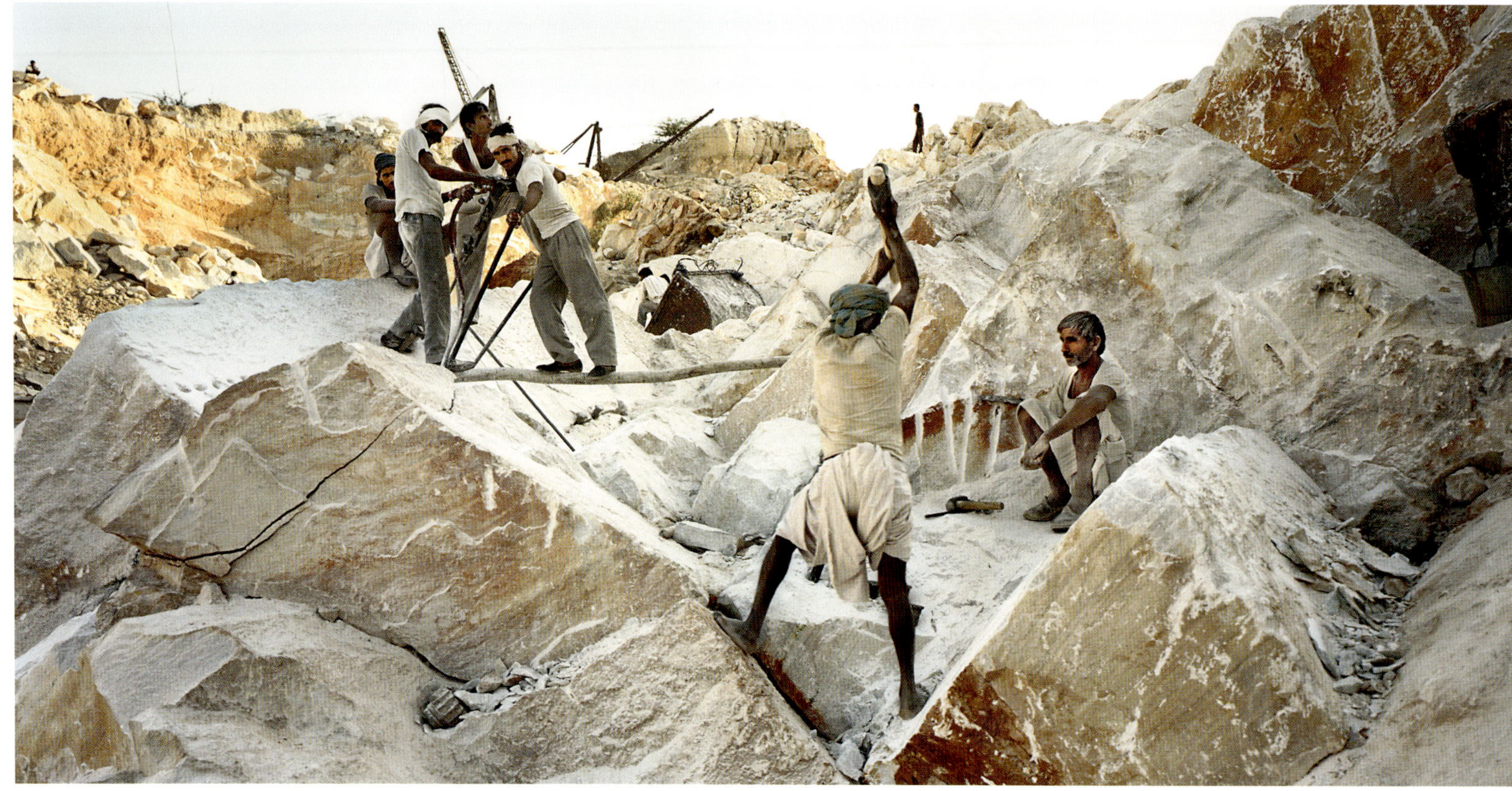

The marble quarries of Makrana have supplied stone for some of the most storied and beautiful structures in all of human history — the Jain temples of Mount Abu and Ranakur and that shimmering monument to lost love, Agra's Taj Mahal. However, for many Indians the Makrana quarries are the Pits of Death. At least half a dozen workers die in the quarries every month and there have been more than 500 serious accidents in the past half decade. Additionally, there is an enormous and poorly documented host of occupational illnesses and injuries— silicosis, deafness and loss of limbs. In 2005 alone, more than fifty local mines collapsed. This beauty has an enormous capital cost.

As with everything in India, the numbers boggle the mind. During the mid-1600s more than twenty thousand artisans and workers laboured to complete the Taj Mahal. A thousand elephants dragged marble blocks some two hundred miles from Makrana's quarries to Agra, arriving at the construction site by climbing a man-made ramp through Agra that was over ten miles long. Beginning in 1631, the town witnessed an endless parade of these elephants that would last for twenty-two years. India's "Monument of Love" is also a monument to much human suffering.

And it continues. There are more than eight hundred marble quarries at Makrana today, varying in size from fifty feet square to over a thousand feet long, the average being only about one hundred by two hundred feet. Most of these mines are illegal. They are worked by *Dalits*, people of the lowest caste, who are given neither safety shoes nor hard hats nor any protection against the clouds of heavy dust that coat their lungs. The pair of tiny barefoot figures hand-over-handing by cable in the lower left on page 115 are descending via the only worksite transport provided at Makrana. The image on page 109 highlights the continuing local reliance on the obsolete channeling system to cut blocks of Makrana marble. The requisite dynamite blasts are regularly set off with no advance warning to workers in the pits. A worker here must be fearless. There is always an adjacent mosque for prayers by the Muslim workforce (page 115). When Burtynsky was photographing the region's largest quarry (page 111), an enormous block being hoisted by cable slipped free and crashed down the bedding slope, firing rock chips at the photographer and his crew. For the quarry owners, workers who survive to age thirty-five are old and obsolete.

Most quarrymen and women are bonded labourers. When they die or are forced to retire, their debts — and hence their bondage — are passed on to their children, beginning the cycle of near-slavery anew. These child labourers can be as young as ten. While they struggle with heavy loads alongside their parents, their younger siblings play on the unstable pit rims high above their heads.

Along with Makrana's human toll come horrific environmental costs as well. Rajasthan is one of India's driest states. The vast quantities of water used to cool cutting saws and contain the stone dust have lowered its water table by six or seven metres in just the past five years. Much of the remaining water is so contaminated with stone slurry that it is effectively useless for drinking or sanitation. Where India's national Highway 8 cuts through Rajasthan between Udaipur and Gomti Chouraha, mountains of marble waste (some sixty-five percent of all the rock quarried is discarded) line the roadway for ninety kilometres. Farmlands have been destroyed and forests of marble blocks, as well as the ubiquitous slurry, have replaced all natural vegetation.

None of this seems likely to improve soon. Government corruption is pervasive; false licenses and fake inspections common. And the economic pressures only increase month by month. Rajasthan is home to ninety-one percent of India's marble reserves and sixty-five percent of all of its natural stone production. Dimensional stone work in Makrana provides employment for some fifteen thousand people there and for more than a million people in all of India. But these jobs are no longer guaranteed. China, that other population colossus, recently surpassed India as a global supplier of stone.

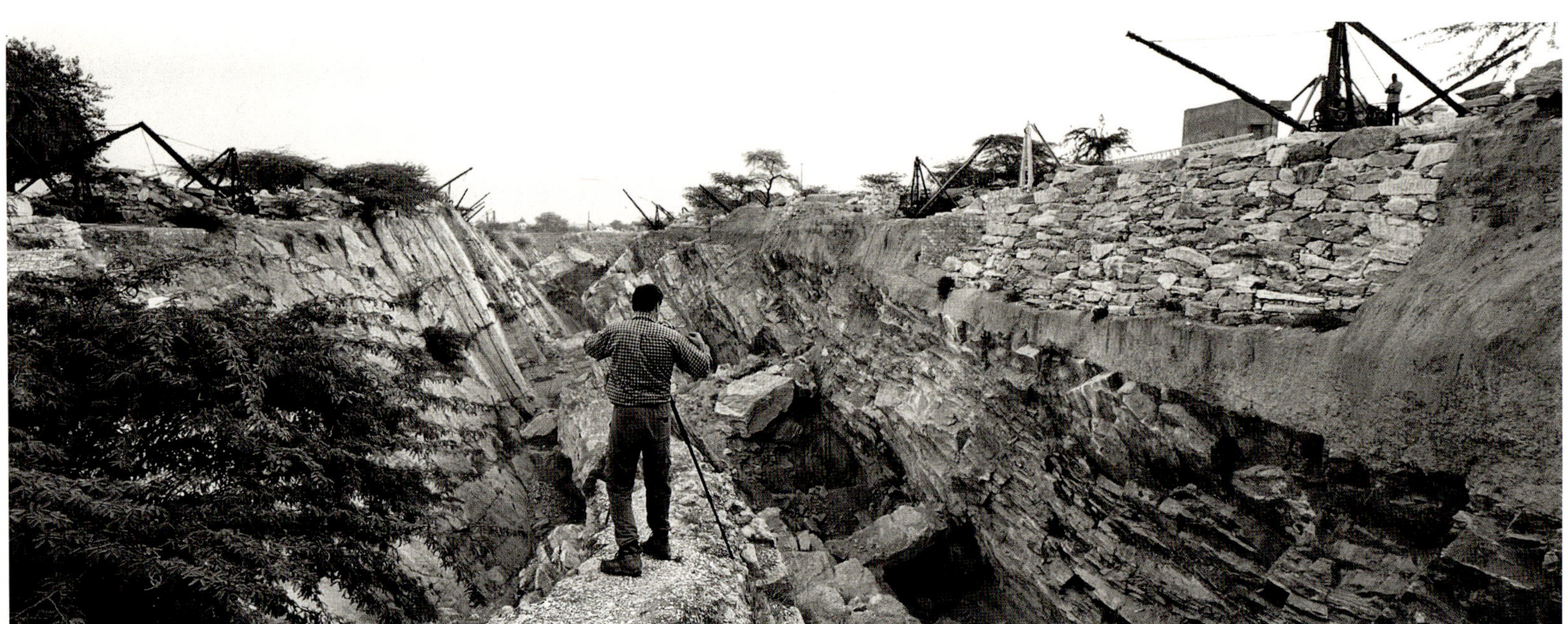

XIAMEN - BUY LOW, SELL LOW

In the year 987, a young woman named Lin Mo left home to climb Mount Meifeng in southern China's Fujian province. When she reached the summit, a strange cloud appeared, accompanied by mysterious music. Lin Mo leapt into the cloud and the winds carried her away. Although her family never saw her again, in the ensuing years she would often appear magically, dressed in flaming red robes and flying off the Fujian coast to protect ships and sailors on the seas between China and Taiwan. The local people called her Mazu — Ancestral Mother — and began worshipping her as the guardian angel of sailors and commoners. Today, wherever Chinese people have gone in the world there are Mazu temples. She has one hundred million devotees.

Mazu has had no shortage of ships and sailors to protect in Fujian. During the Ming Dynasty, the region's greatest city, Xiamen, became an important base for battling pirates. In the mid-1500s it became China's first port for European trade, and at the end of the first Opium War between China and Britain, it was one of the five Chinese treaty ports opened in 1842 by the Treaty of Nanjing. During the 19th century, Xiamen's main export was tea. (Pekoe comes from the local Amoy dialect, as do the words satay and ketchup.) Today, many of the ships clearing the harbour at Xiamen under Mazu's protection are loaded with goods made of local granite and imported marble. China — and Xiamen in particular — has become a major player in global trade in natural stone. In 2004, China exported nearly five million tons of stone products. Japan was the largest customer.

Since Deng Xiaoping began opening China's economy to the world in 1978, economic growth in coastal provinces like Fujian has been phenomenal. The high-grade granites near Xiamen have led to the development of more than a hundred local quarries and thousands of stone product workshops employing nearly a million people. The skilled stone carvers seen working on page 139 will make you a life-size likeness of Michelangelo's David in granite for under a thousand dollars. Send them a photograph of yourself and they'll do one of you for the same price.

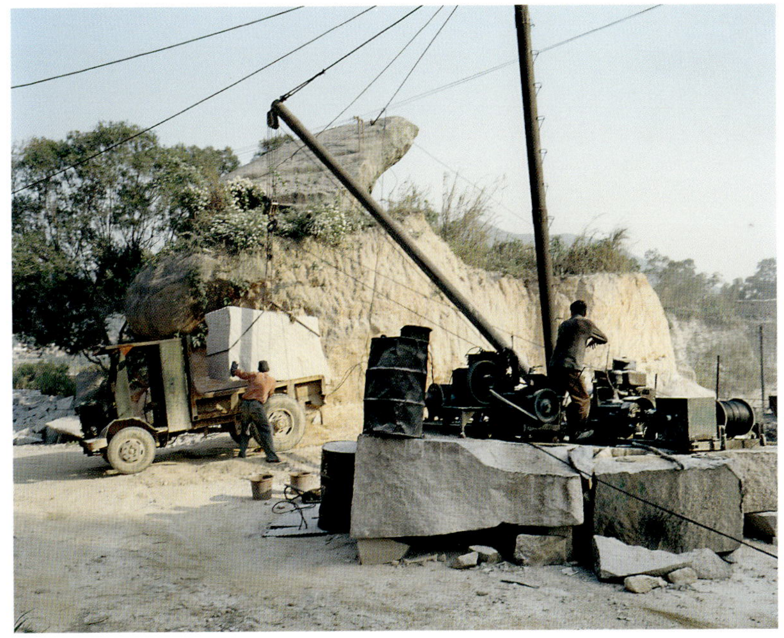

The story of this trade in China is very similar to that of its main rival, India. Their vigorous assault on the land reminds us of our own experience in the West more than a century before. The pressure on China's land is intense and conflicting. As whole mountains are levelled and crushed for gravel to build cities, traditional farmers are squeezed onto ever-smaller islands between quarries and their access roads (page 131). There is much waste, pollution and occupational disease. No safety equipment or masks are employed. (The whole sad story of the Barre, Vermont, stone-shed workers of the early 20th century will soon be reprised at Xiamen in the early 21st.) Quarry machinery is basically as primitive as that used at Makrana and elsewhere in India. Skilled Chinese stoneworkers in the interior are paid only $2.50 for twelve-hour days. Some are as young as ten.

But none of this slows China's new breed of aggressive entrepreneurs. The 2005 Xiamen International Stone Fair had twenty-two hundred booths. In 2006, buyers came from twenty-five different countries, and the locals' banners and web sites shouted at them at the top of their lungs: "Sell Roofing Slate, Sell Stone Countertop, Sell Bathroom Sink, Sell Fountain, Sell Granite Baluster, Sell Granite Flowerpot, Sell Stone Bench, Sell Granite Ball, Sell Columns, Sell Abstract Sculpture, Sell Gravestone." Local companies promise "very favourable price and high quality." The owners will meet you at the airport. As one company in the local trade association bragged on its web site, "As all are aware, Chinese are capable of making anything. They are like bees who would not be idle."

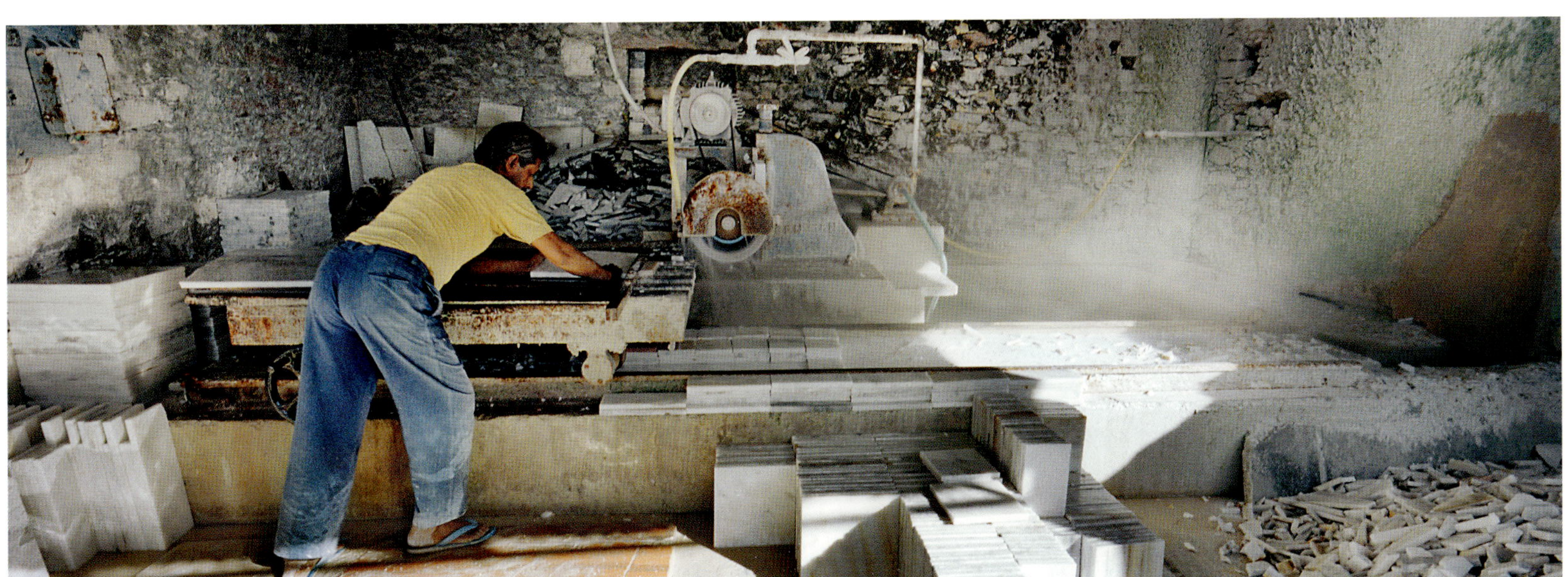

IBERIA - INVERTED ARCHITECTURE

In the 36th volume of his encyclopedic *Historia Naturalis*, the Roman scholar Pliny the Elder describes the Iberian Peninsula as being "rich in marble, both white and coloured, of excellent quality...." He knew what he was talking about from first-hand experience. The emperor Nero had made Pliny his procurator, a sort of managing agent, for Spain during the second half of the first century. By his time, the Romans had been quarrying in both Spain and Portugal for decades. The *Teatro Romano*, built by the Roman emperor Agrippa at the western Spanish city of Mérida in 16 B.C., was constructed of marble brought in from Portugal. It is still used today. The elaborately decorated column capitals that survive on the Roman Temple of Diana in the ancient walled city of Evora, southeast of Lisbon, were also carved from local marble. Portugal produces easily worked, fine-grained marbles of great durability.

In Spain, marble quarrying at Macael, in Andalucia, goes back to at least the Phoenician era five hundred years before Christ. There are early 1st-century tombstones at Tíjola in Almería carved from Macael marble and numerous 3rd-century marble columns, capitals and statues at Italica, near Mérida, that have been traced to Macael quarries. The Romans obviously preferred using cheaper local marbles rather than importing expensive stone from their imperial quarries back home.

However, it was after Arab colonization in the 700s that an organized marble industry began in Spain. The Arab colonizers loved Spanish marble. In addition to wall and floor tiles, they crafted many fountains, pillars, stairs and baths from it. The five hundred columns of the huge mosque in Córdoba, dating from 780 A.D., are Andalusian marble, as is much of the stone in Granada's 13th-century Alhambra. In 1501, after the re-conquest of Spain from Arab rule, the Catholic monarchy of Isabella and Ferdinand granted the Macael municipality the rights to quarry exploitation. Over the next three hundred years, ecclesiastical buildings and palaces constructed of Macael marble would include, among many others, the cathedrals of Jaén and Seville and the Palace of Charles V in Granada, as well as the royal palace in Madrid. Five hundred years after the monarchy gave it quarrying rights, Macael still owns all the quarries as a municipality. The two hundred marble companies there, often family-owned, currently operate under five-year leases and employ about 5,500 people. At one enormous quarry in Alicante (page 156 - gatefold), Burtynsky was frustrated to discover there were over twenty different family operators, making obtaining permission to enter a logistical impossibility. However, once again an obstacle proved to be an opportunity. The distant view, with the grand panoramic sweep foregrounded by olive groves and fruit trees, proved much more telling than detailed views of sections of quarry faces. It encouraged the photographer to experiment and create his first triptych in the quarries series.

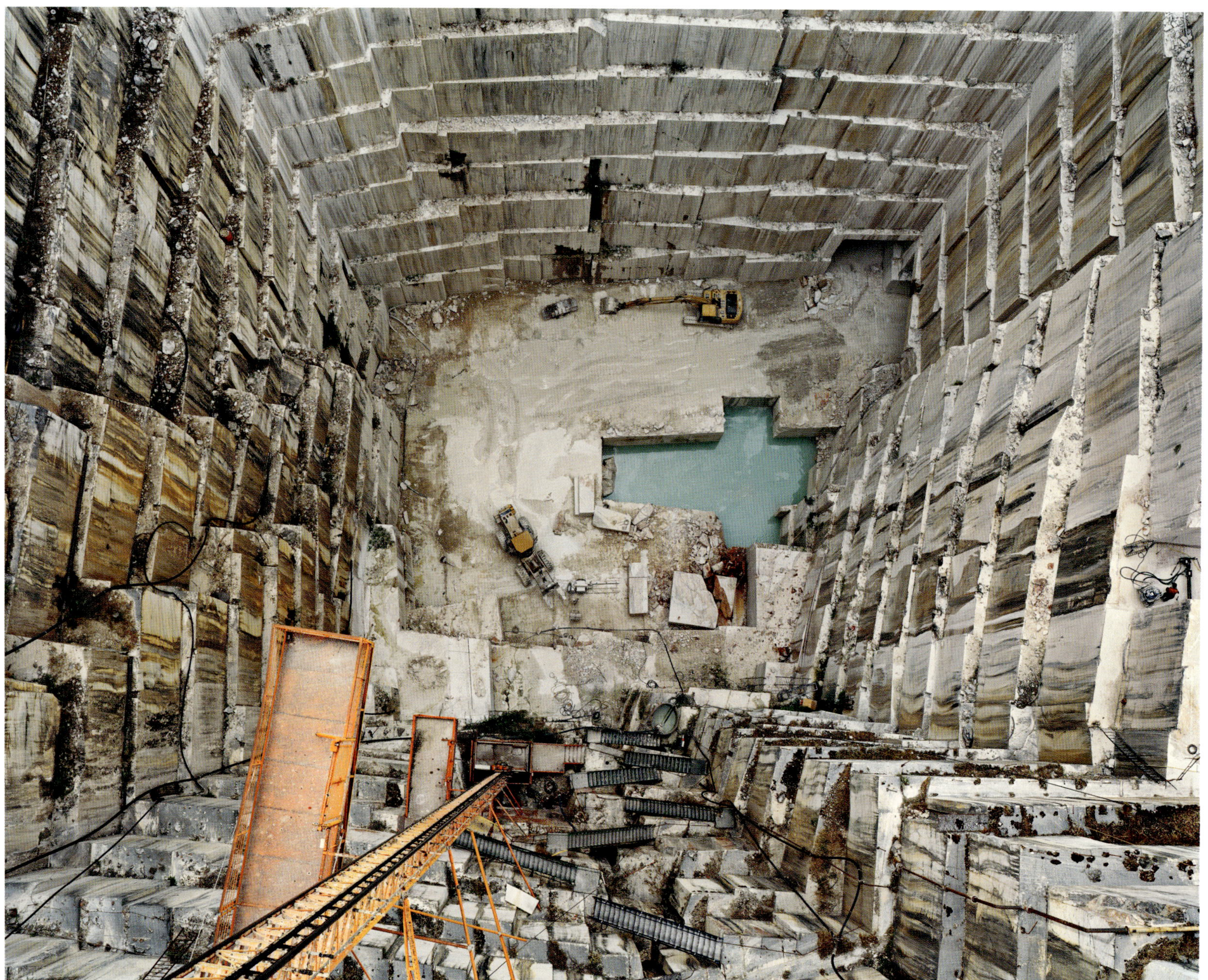

Iberia Quarries #8, Bencatel, Portugal, 2006

Until the early 20th century, Spanish quarries used the same methods — hand cutting and transport by oxen — as the Italians at Carrara had employed since antiquity. In the 1920s, hydraulic saws came into use, and by the 1940s Spanish marble quarries were equipped with electric saws and diesel compressors. Rapid modernization after the 1960s included the introduction of the now ubiquitous diamond saws. Although Spanish and Portuguese quarries now use the same techniques and technology as any other European operation, the works at Macael favour the machinery employed in open-pit coal and ore mines, as their marble is embedded in deep layers between other types of material. Young men from the region who are interested in quarry mining tend to become heavy equipment operators so they can begin earning immediately after high school. Consequently, many of the local mining engineers now are women. They are more willing to spend the extra years studying at technical colleges. Spanish quarry workers are overwhelmingly organized into trade unions.

Although the Macael area is famous for its beautiful and very durable white marble, its stone companies are now making efforts to market the various yellow, gray and green marbles also found in their quarries. Spain is now the second-largest marble producer in the world. While the domestic market is huge, the U.S. is its second-largest market, followed closely by China.

When it comes to marble, Portugal is pink. The quarries of Southeastern Portugal — Vila Viçosa, Pardais and Bencatel — produce Pink Dawn, Borba pink, Estremoz pink and the stone known as Portugal pink. These stones are exported to Spain, France and Germany, while the white marbles these quarries also produce are largely exported to Italy. Portugal is the world's fifth-largest marble producer.

Both countries are now facing the issue of what to do with exhausted quarry sites. Portugal's marble-producing areas are dotted with abandoned pits that are filling up with ominous-looking turquoise water. Repurposing these quarries is not easy where temperatures exceed 50 degrees Celsius by the beginning of summer. In Spain, at Macael, there has been extensive replanting and reforestation around the quarries. This area experiences a very arid climate – think back to the landscapes of the spaghetti westerns that were filmed here. However, the Spanish are working on it, and in one unique case an exhausted quarry has been successfully recycled as a public cultural centre for outdoor musical and theatrical performances.

The quarries of southeastern Portugal are often extremely deep. The operation on page 169 has reached a depth of more than 450 feet. These pits, with their precipitous walls and profound depths, dictated extreme solutions to the photographer's fundamental problem of where to stand. The photograph of the Cochico company's quarry at Pardais (page 151) was made using one of the orange-painted elevator platforms cantilevered from the lower right wall of the pit. Here, nearly a quarter century after his initial encounter with the quartz mine photograph at Canada's National Gallery, Burtynsky had finally found a site that allowed him to make homage to the inspiration provided by August Sander. For Burtynsky, these Portuguese quarries represented the culmination and conclusion of a fifteen-year search for his dream image of an architecture turned inside out and upside down. Turn the image on page 149 upside down and there it finally is – the inverted ziggurat that he had so long imagined. His quarry work was complete.

ENDNOTES:

August Sander, Untitled [Quarry], ca. 1925 – 1935
Gelatine silver print, 16.7 x 25.2 cm.
Copyright: The J. Paul Getty Museum, Los Angeles, CA.
Courtesy: Die Photographische Sammlung/SK Stiftung Kultur – August Sander Archive, Cologne; SODART, 2007

Helicoidal Wire Stone Saw, 1880s
"The process usually (late 19th century) employed in quarries or stone-yards for sawing consists in slowly moving a stone-saw backward and forward, either by
hand or machinery, and with scarcely any pressure. Mr. P. Gray has, however, devised a new process, which is based upon the theoretical considerations given above. His Helicoidal Saw is, in reality, an
endless cable formed by twisting together three steel wires in such a way as to give the spirals quite an elongated pitch."
-Scientific American Supplement, No. 520, December 19, 1885

Taj Mahal, Agra, India
Photo courtesy: Dan Kamminga, 2007

Cararra Satellite Image
Overview represents approximately 22 sq/km.
Copyright 2007 DigitalGlobe Inc., Longmont, Colorado USA

Burtynsky at Makrana, page 22
Makrana, India, November 2000
Copyright 2000 Jim Panou

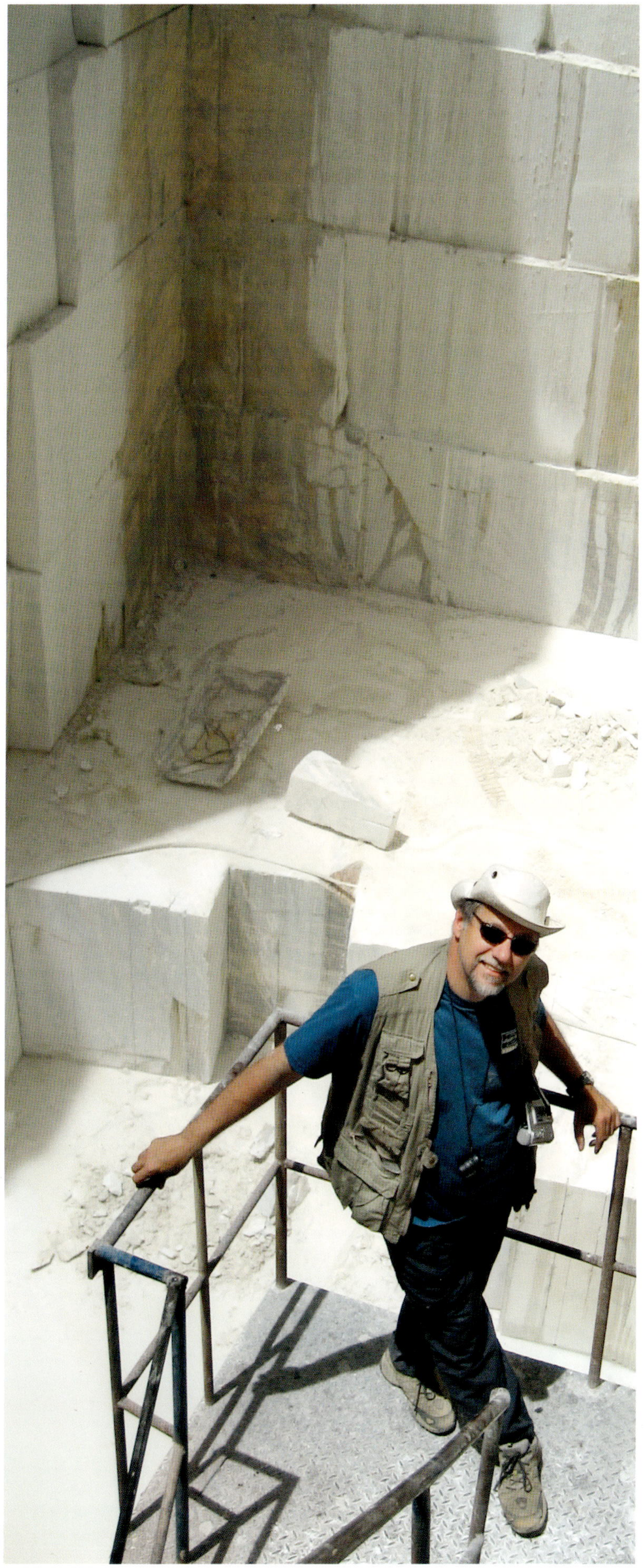

PLATES

ROCK OF AGES

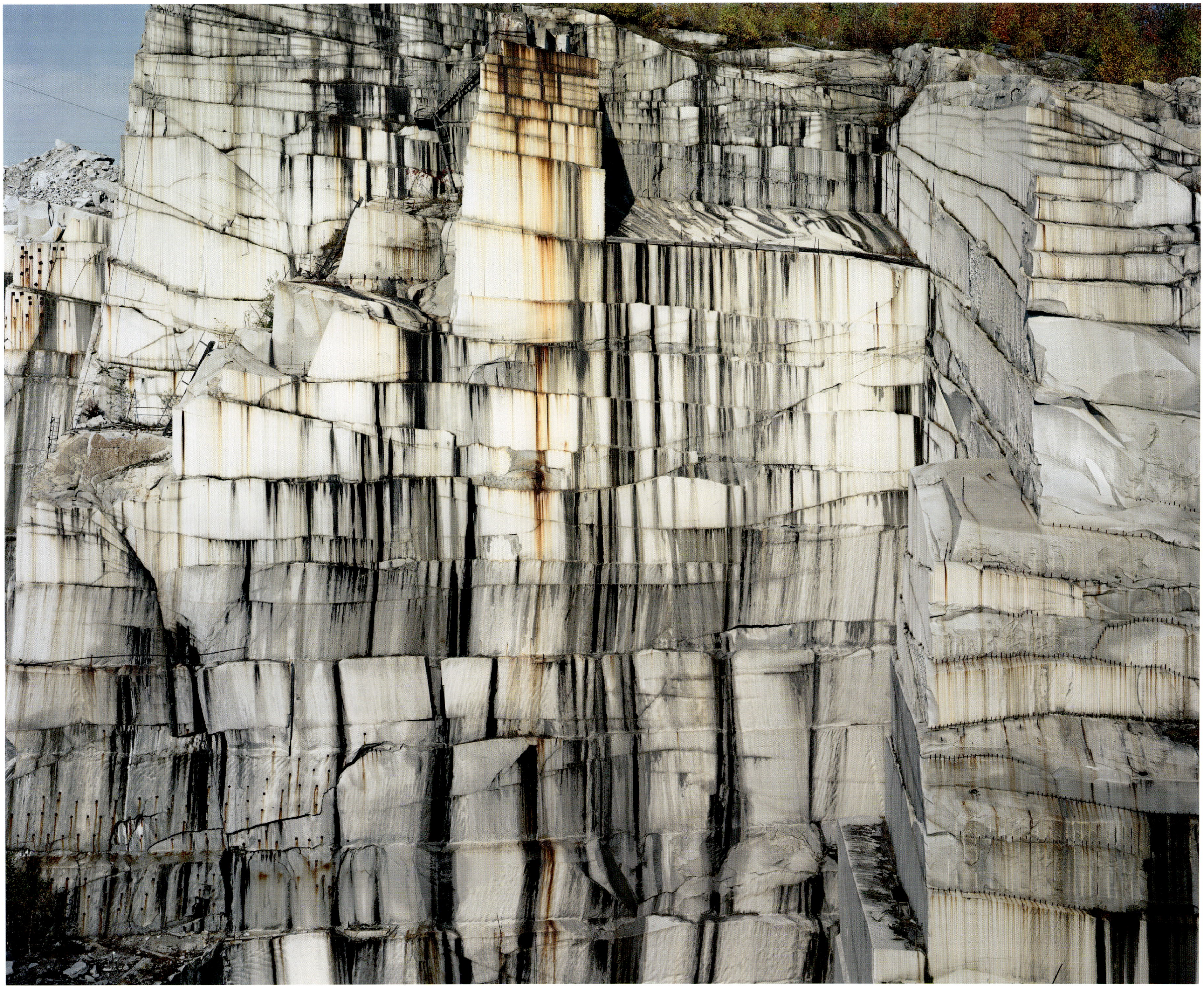

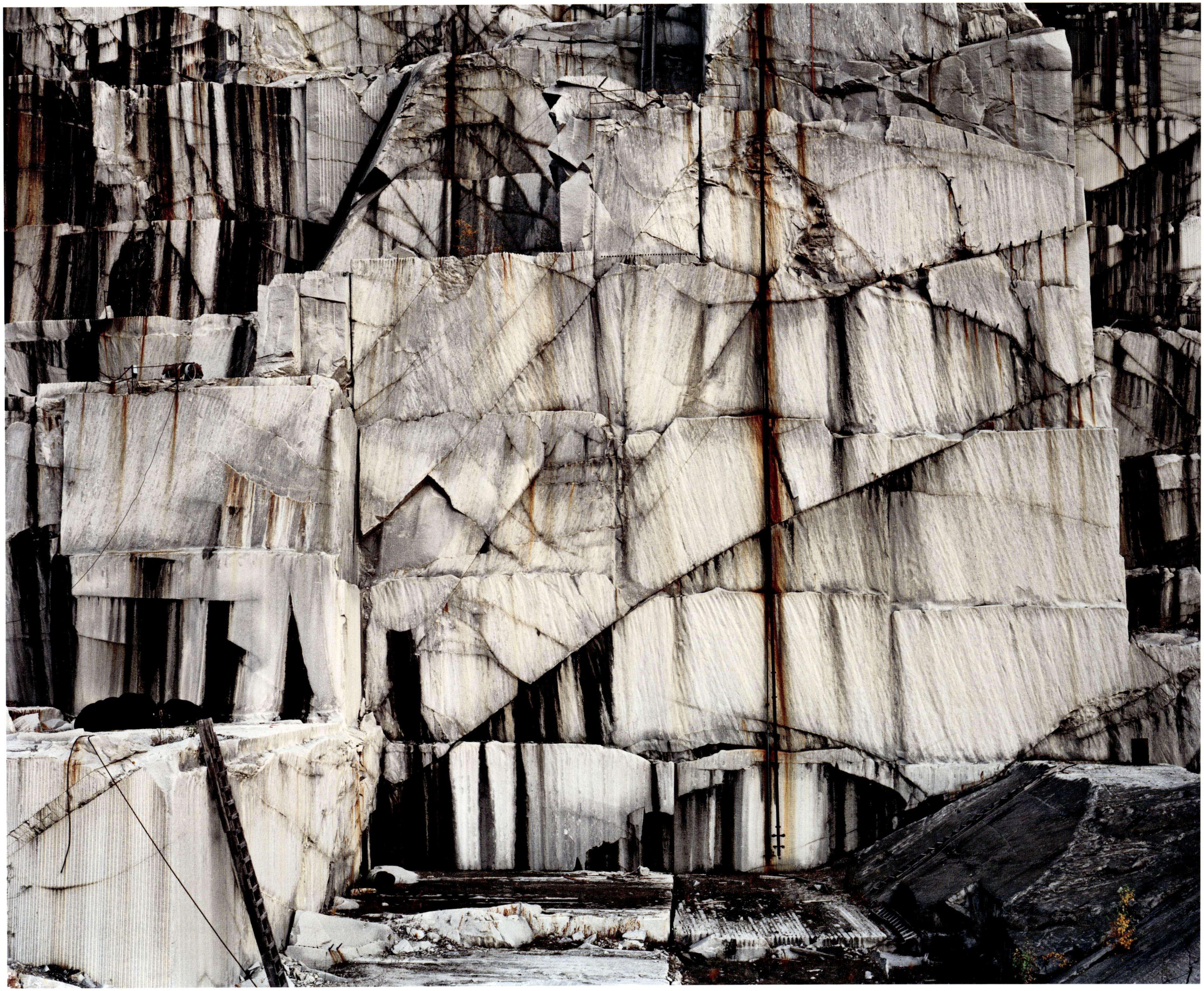

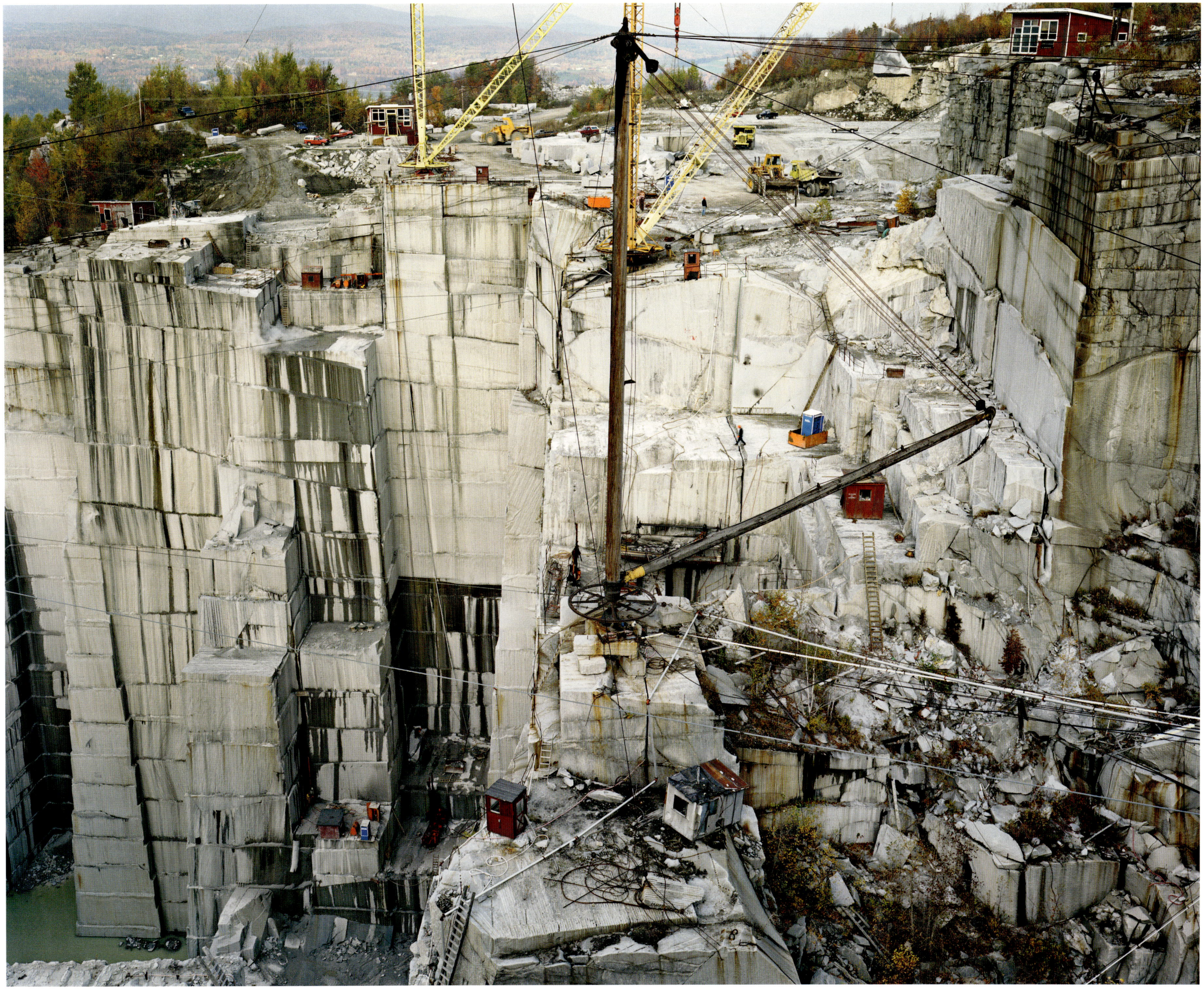

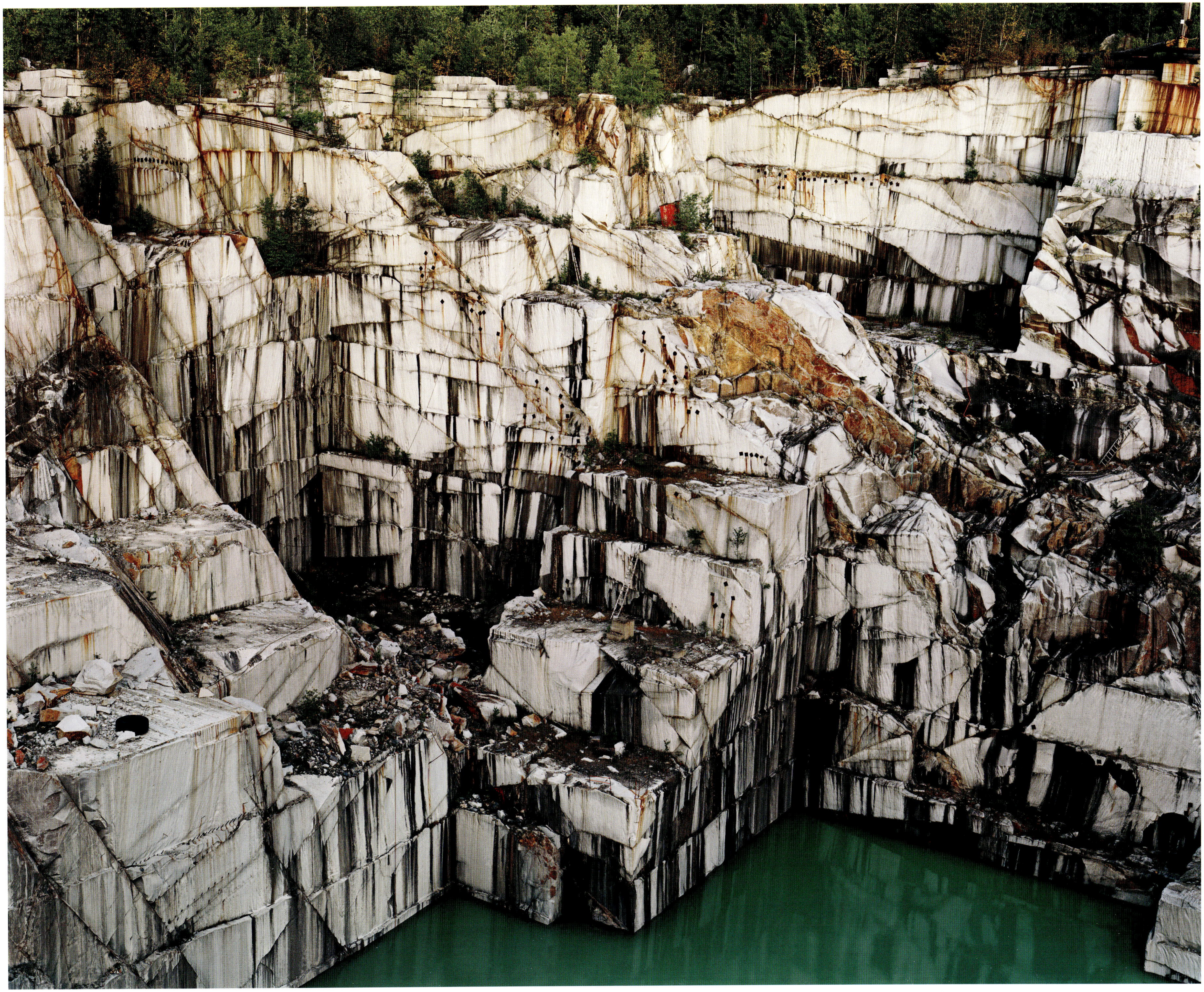

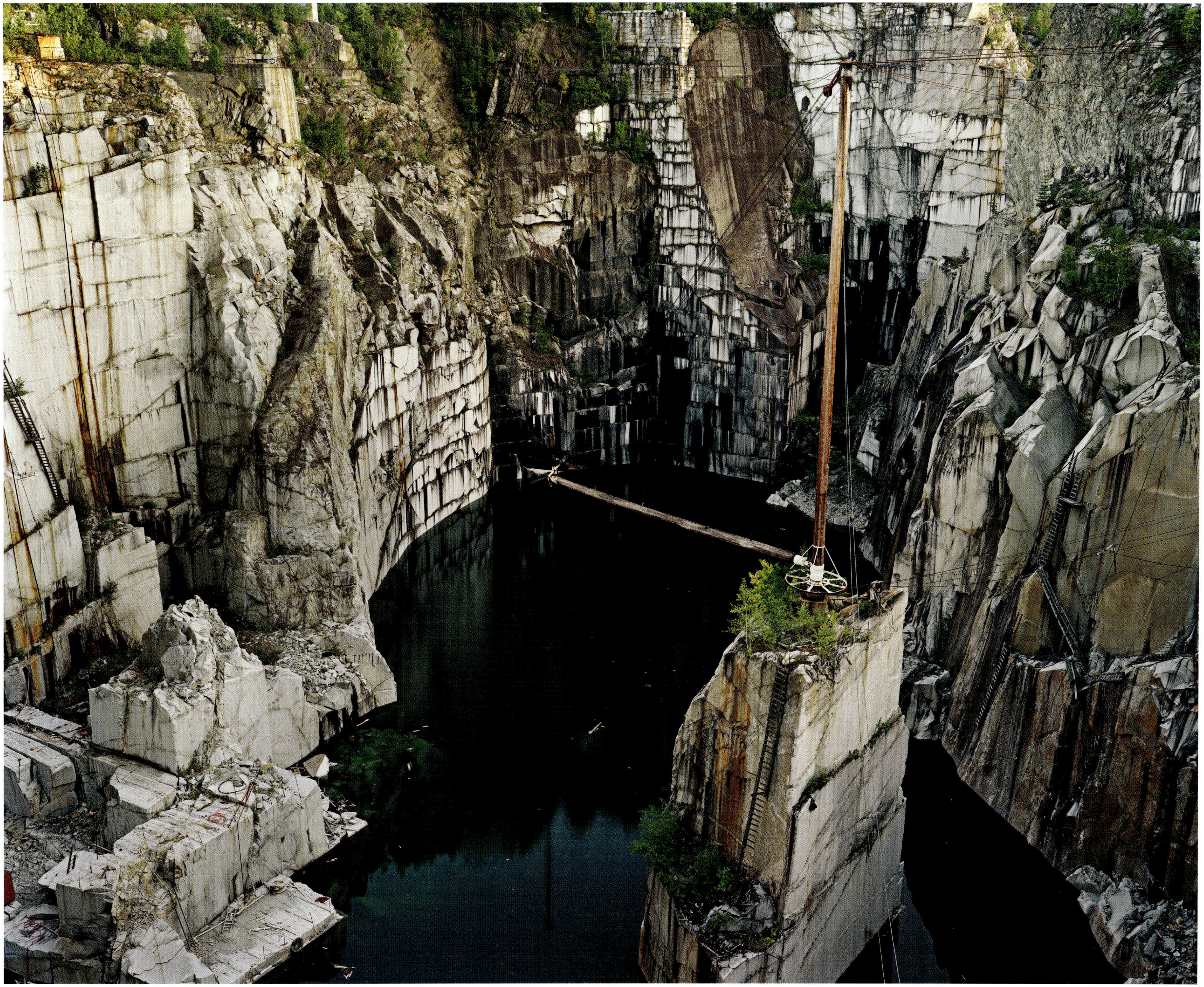

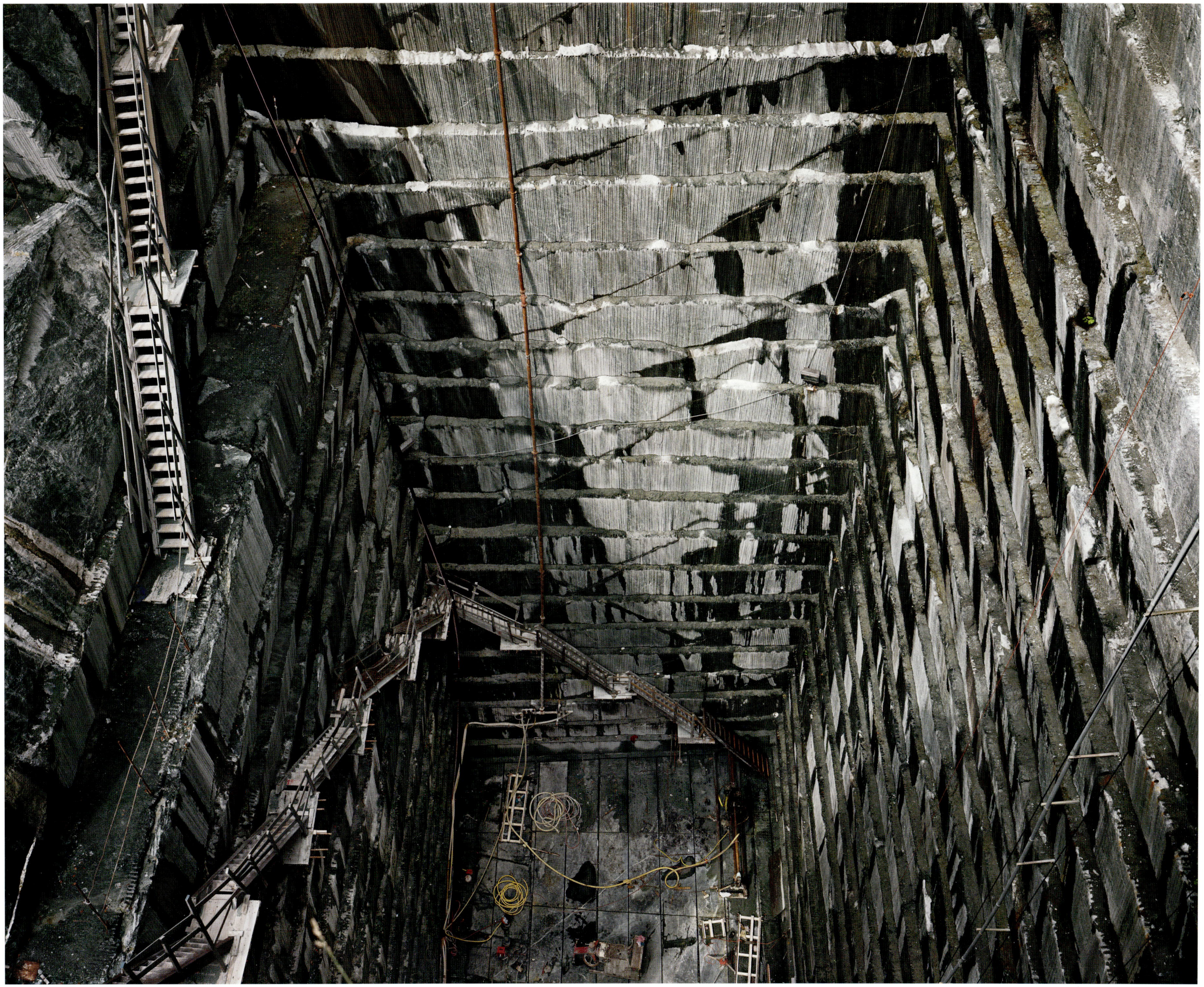

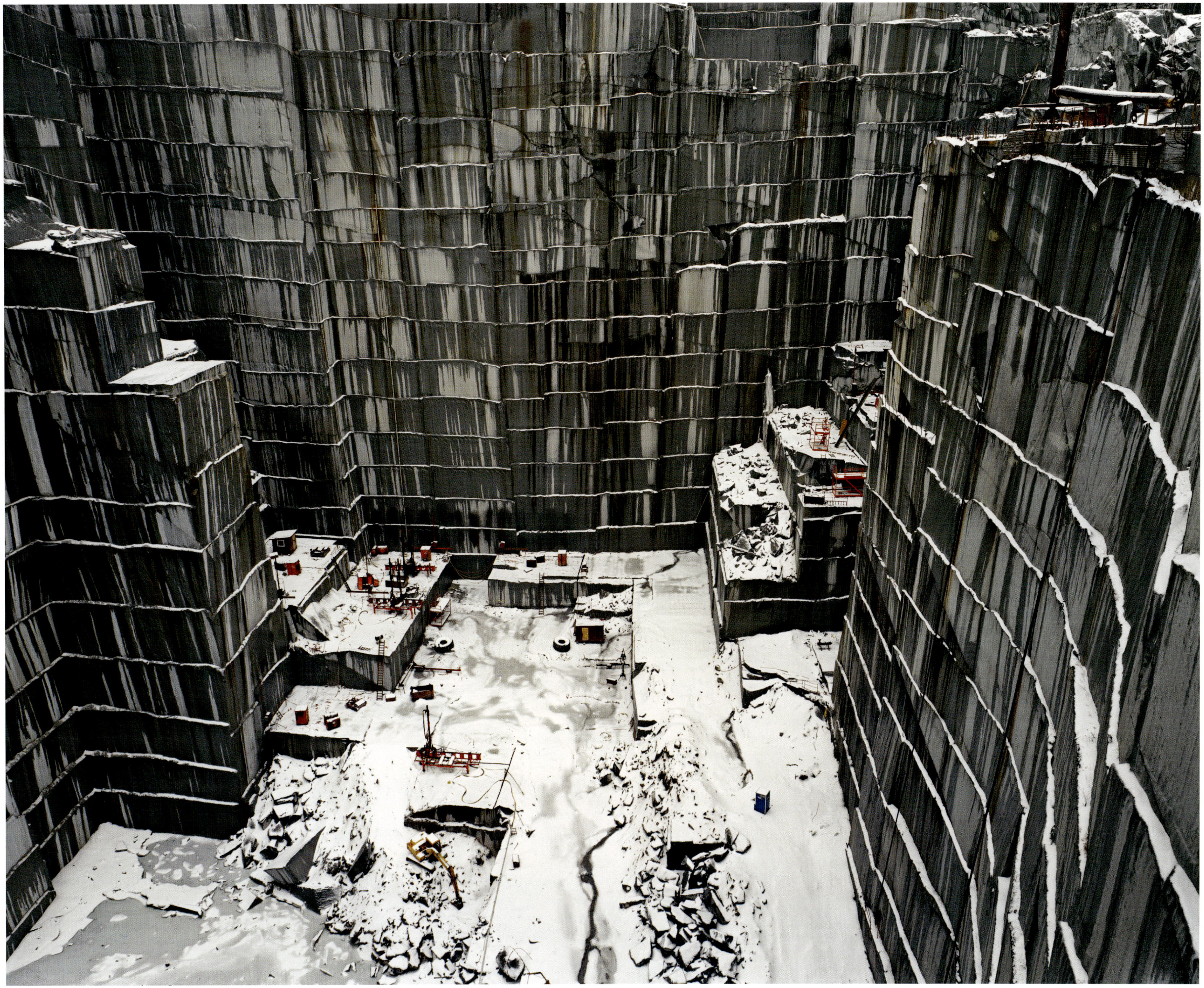

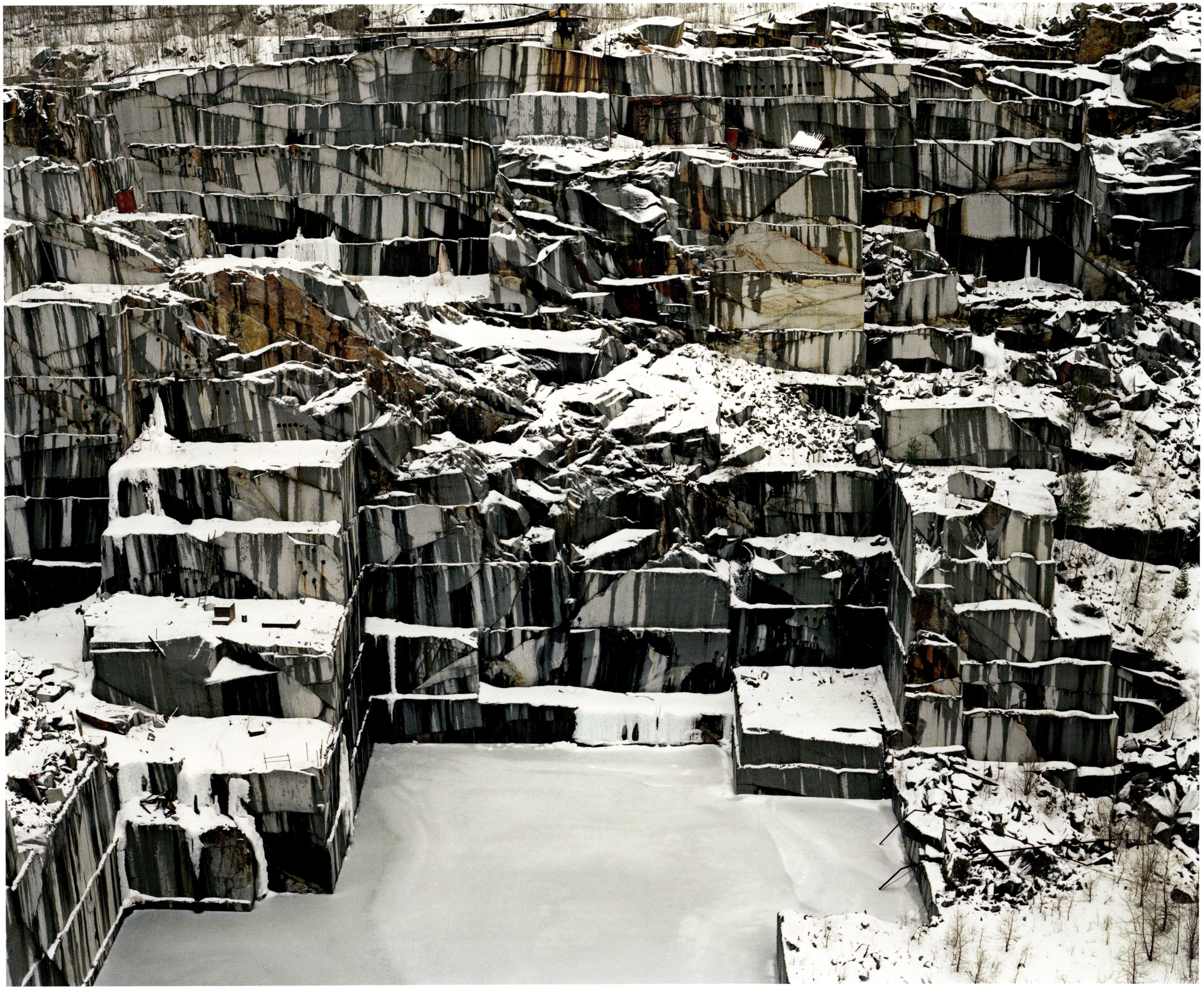

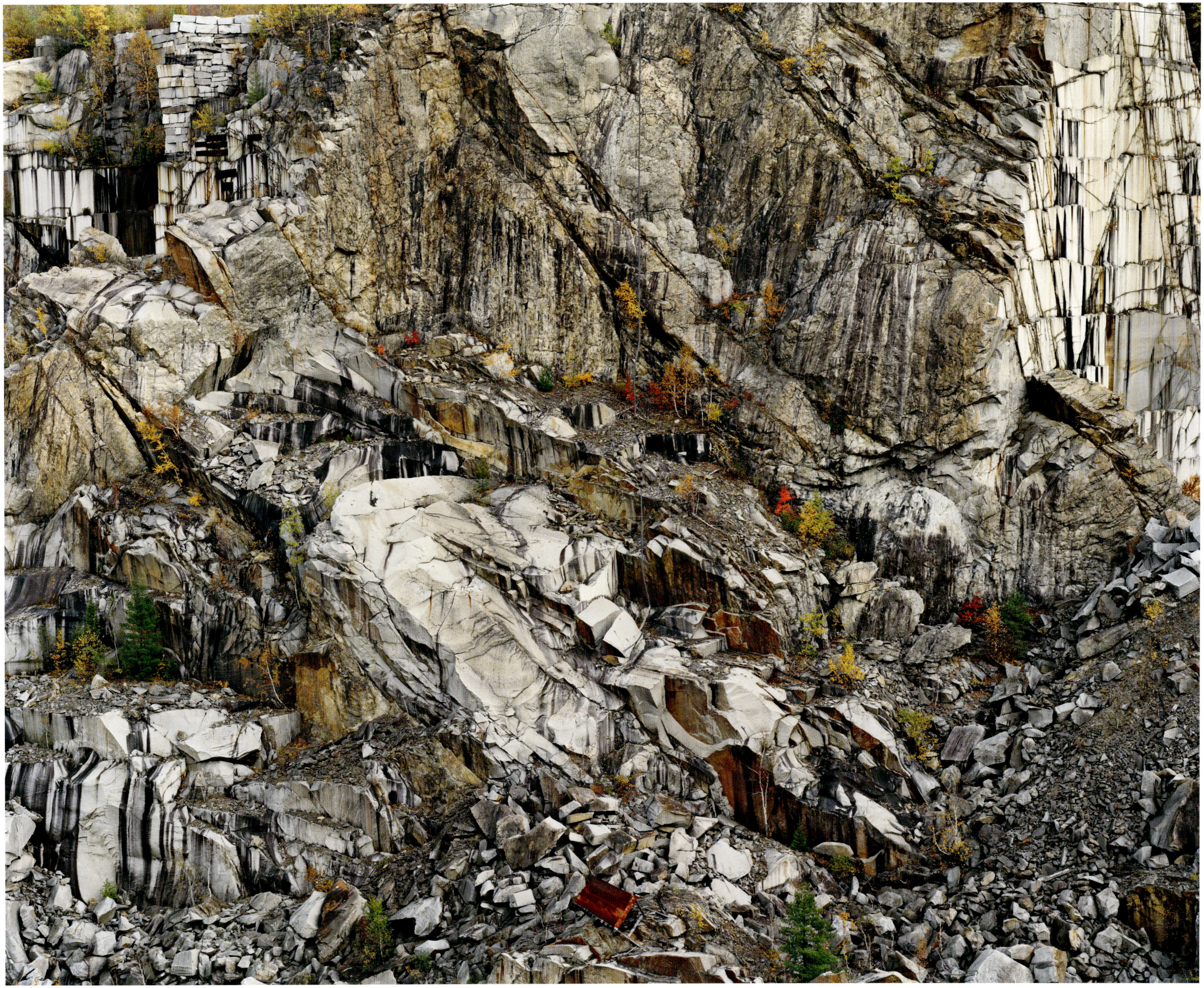

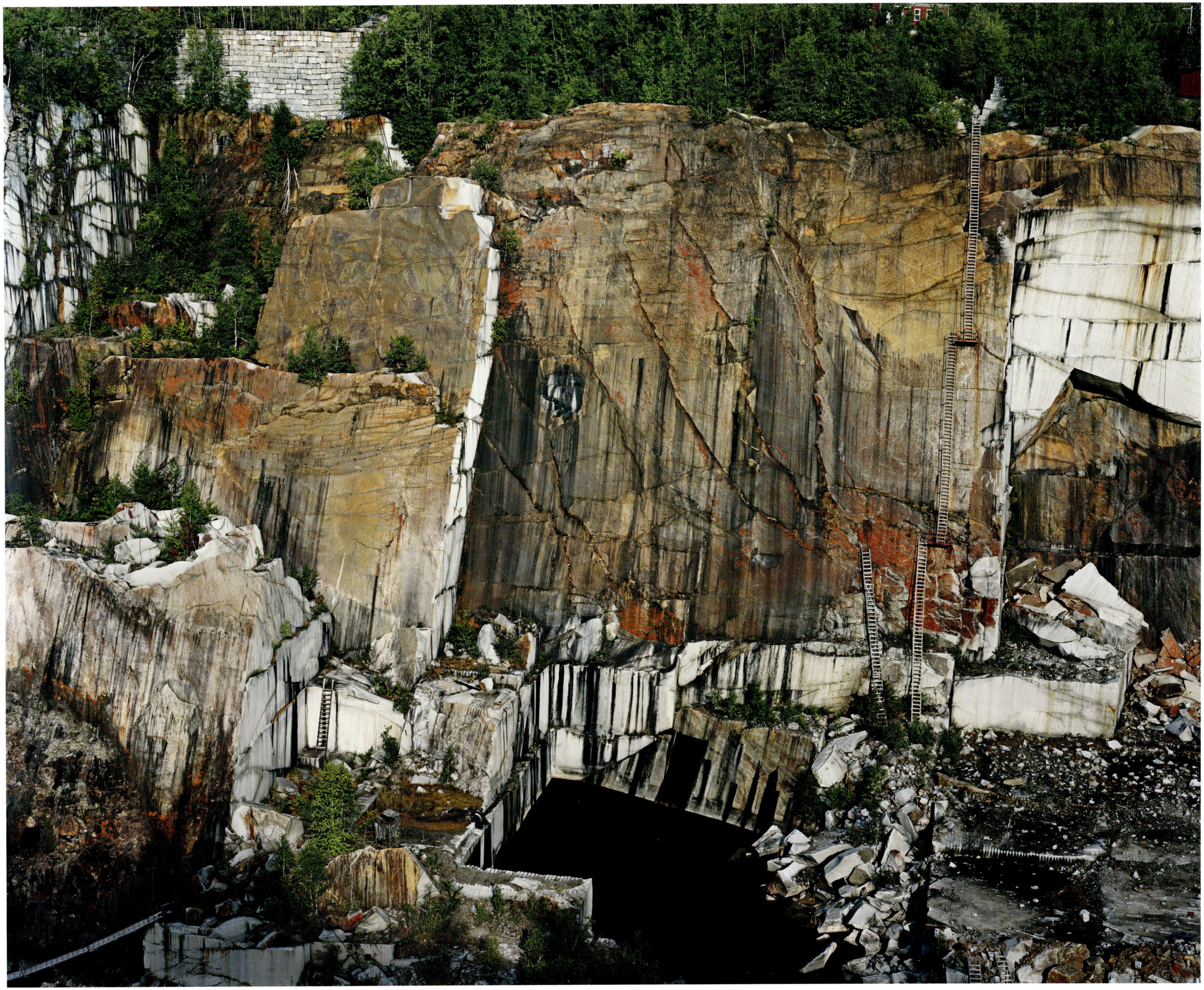

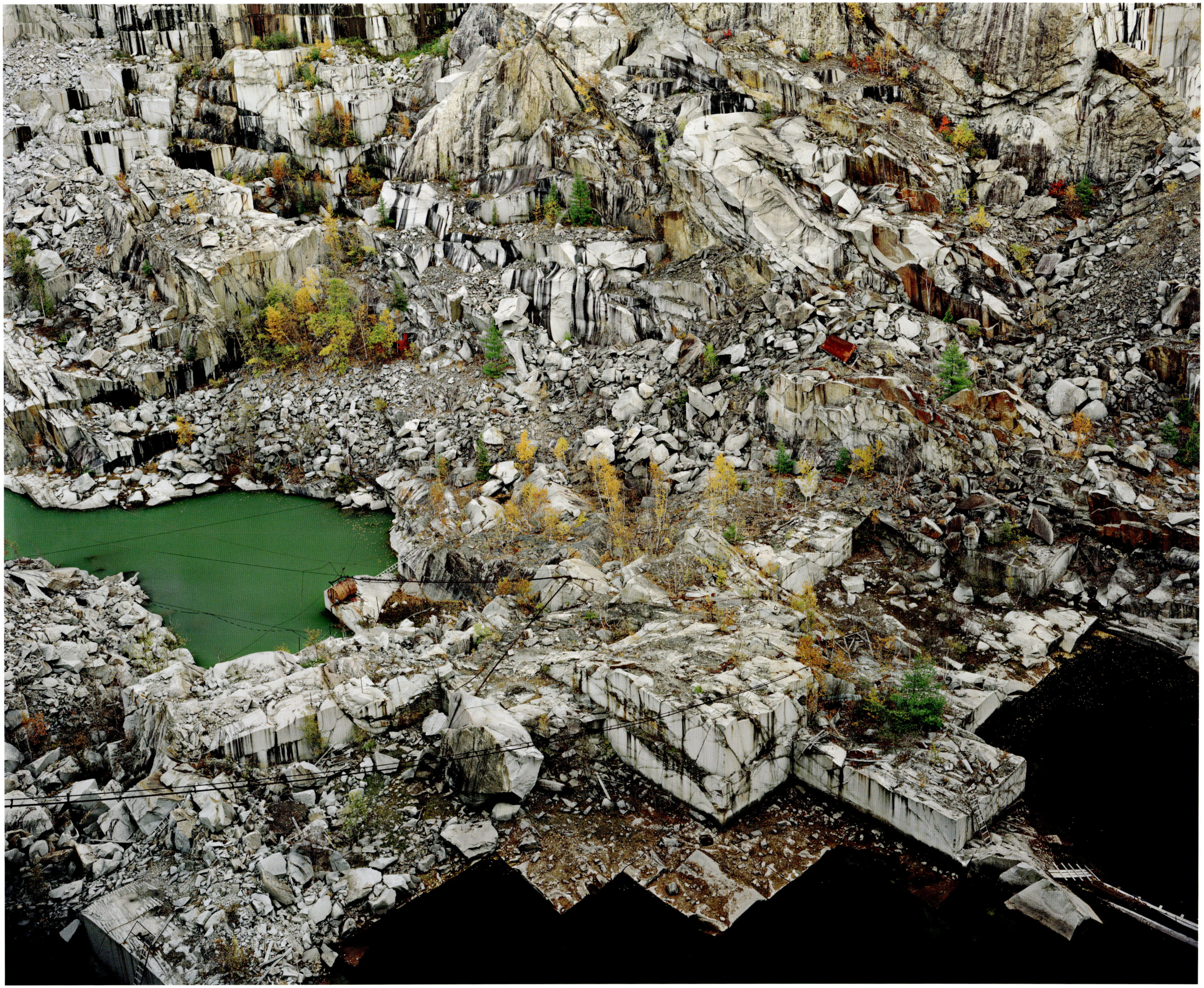

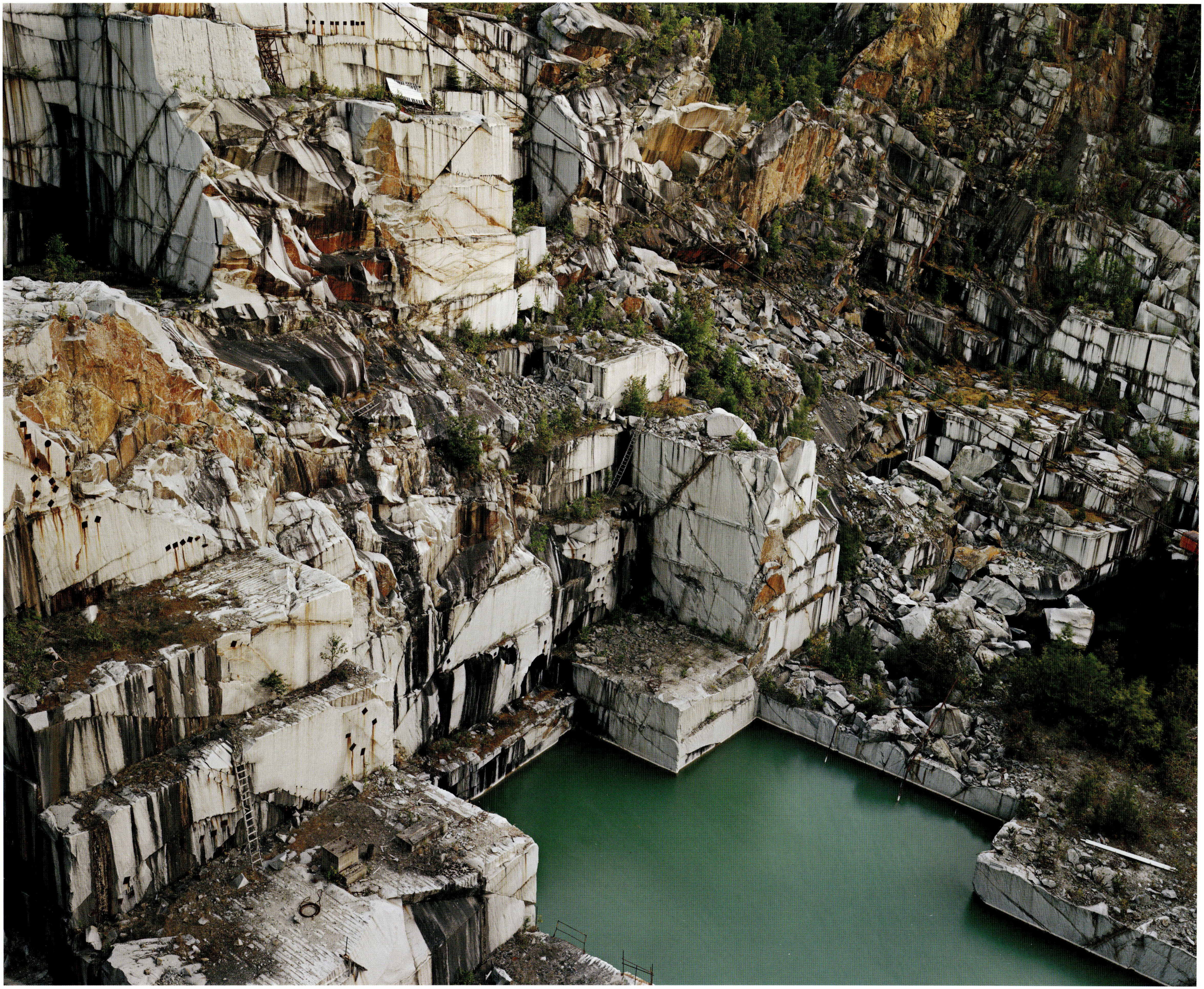

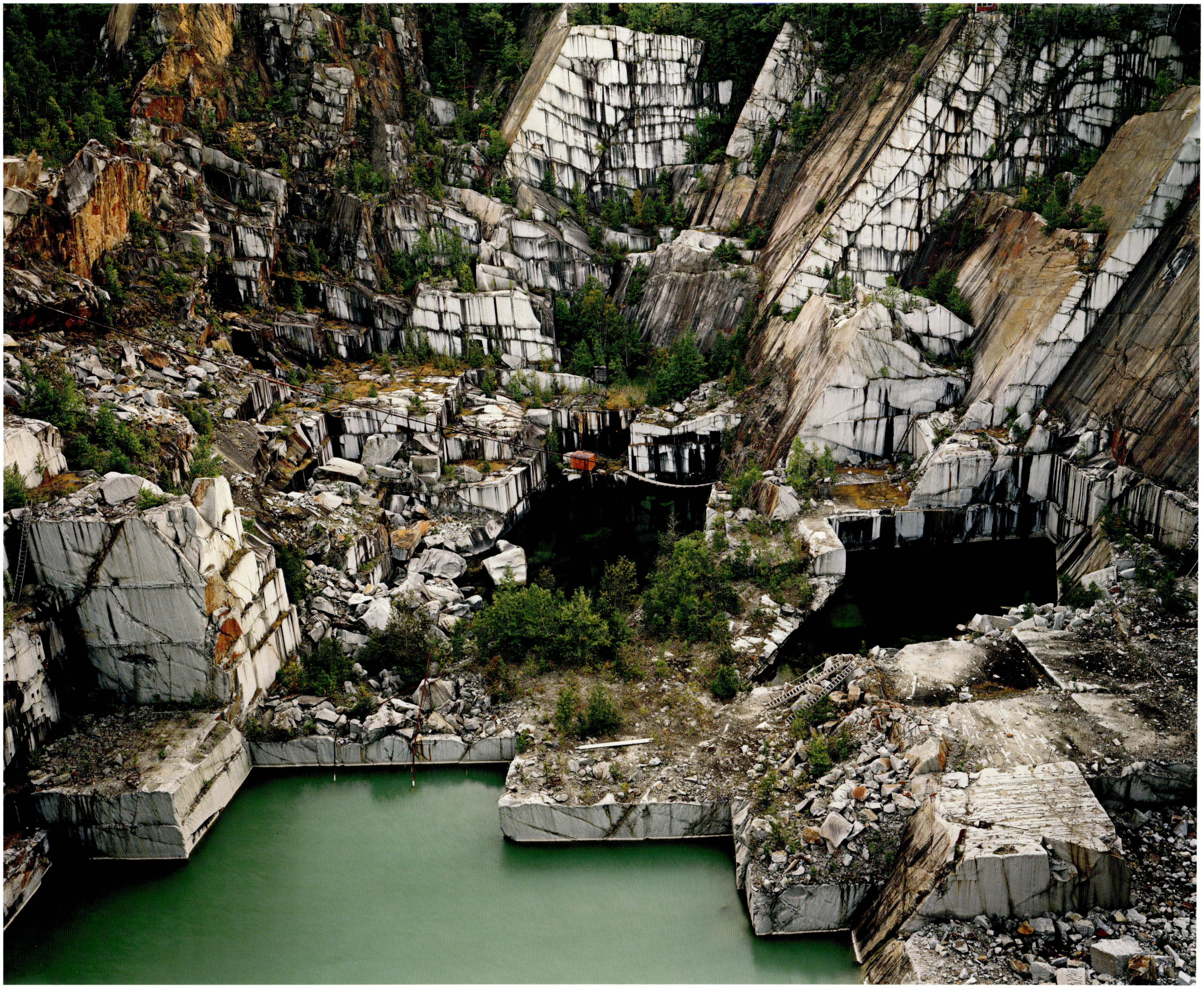

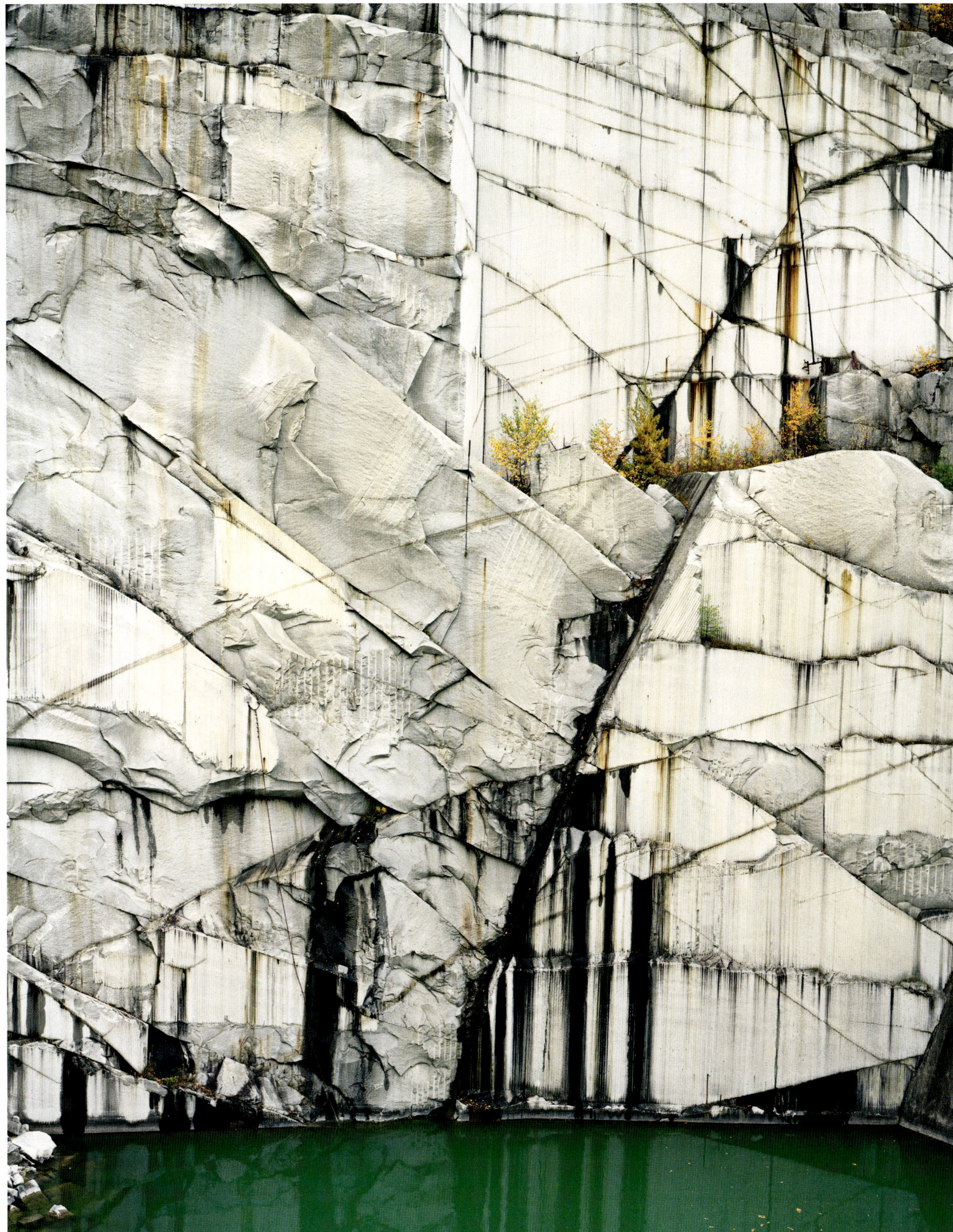

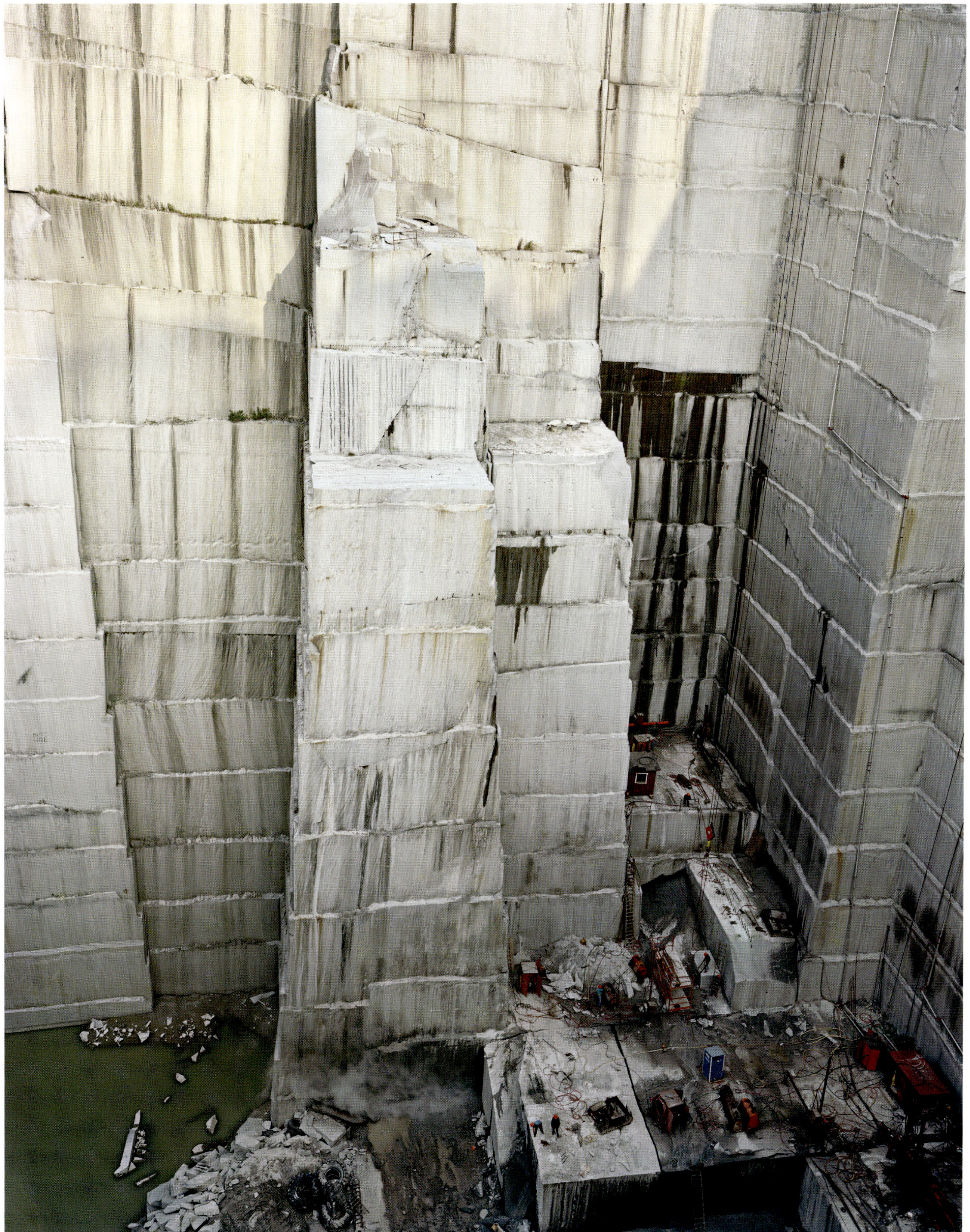

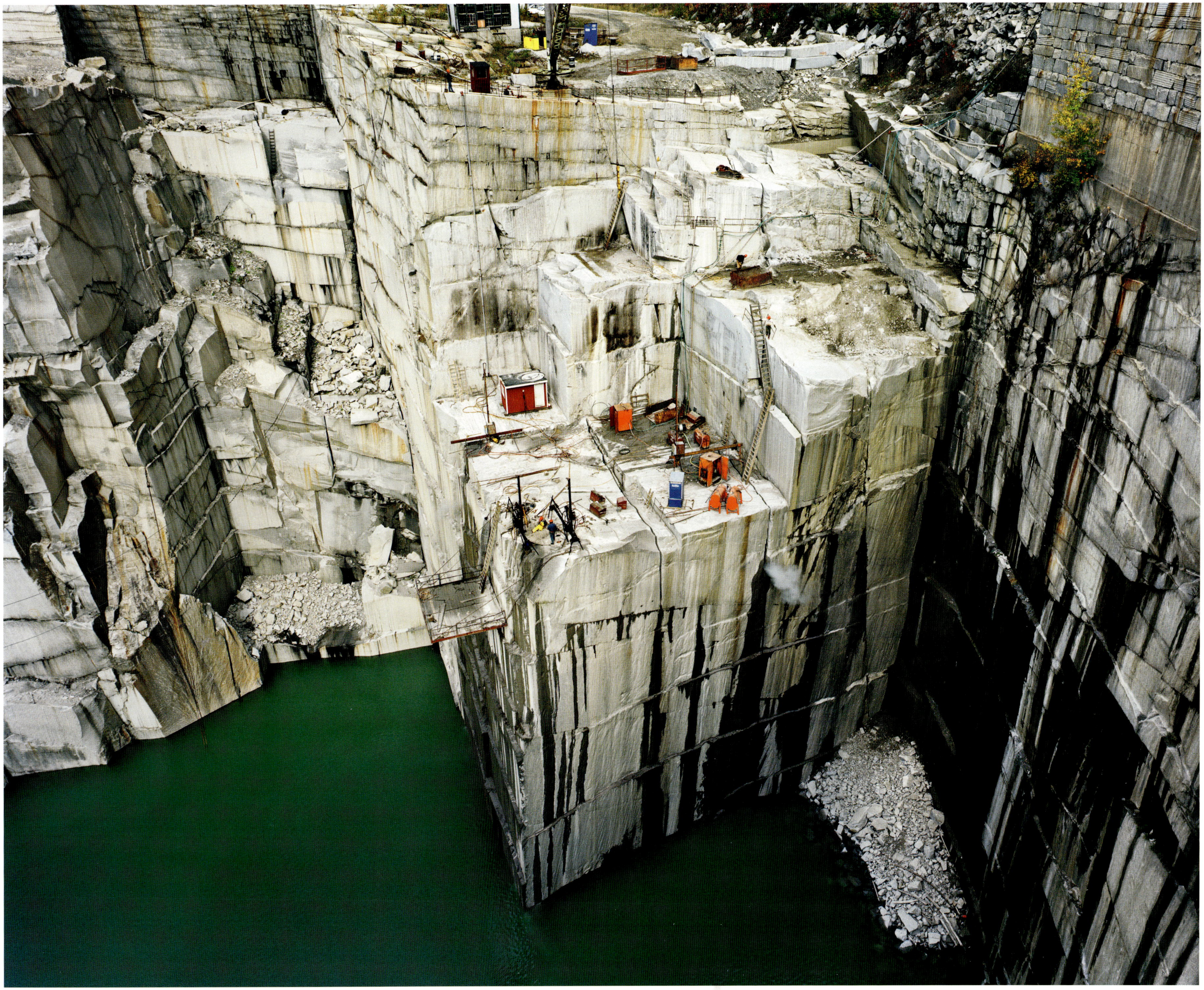

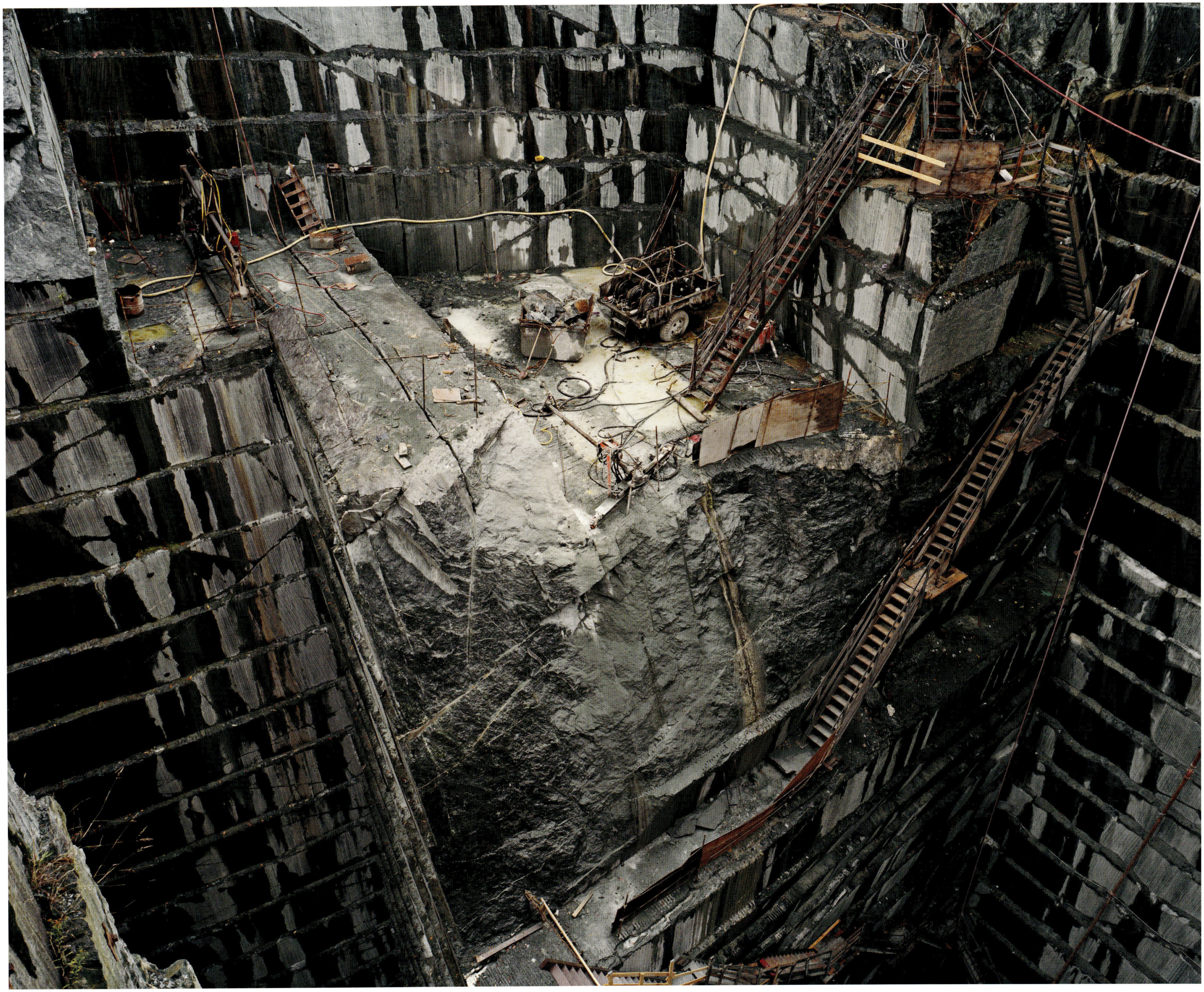

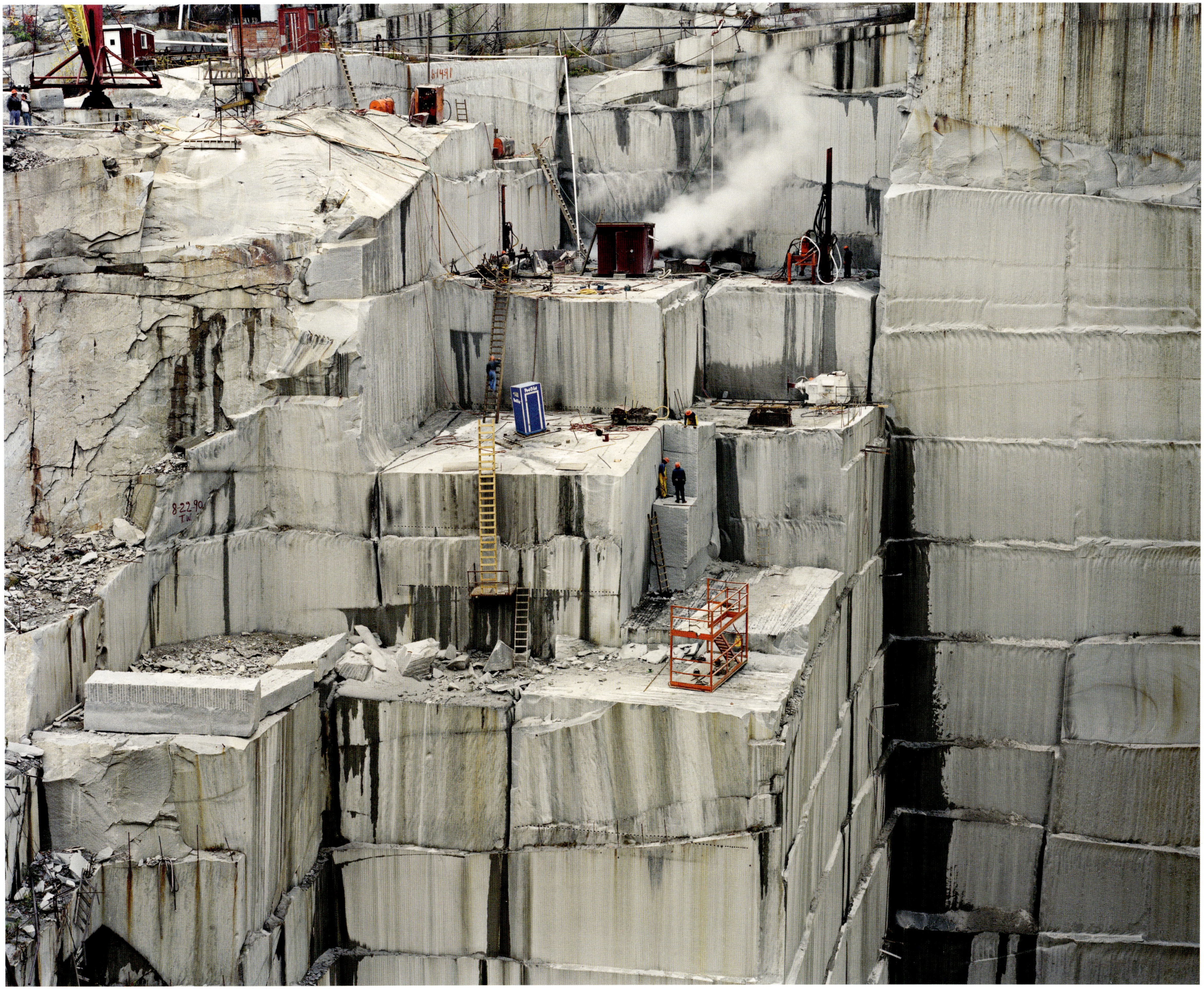
8-22-90

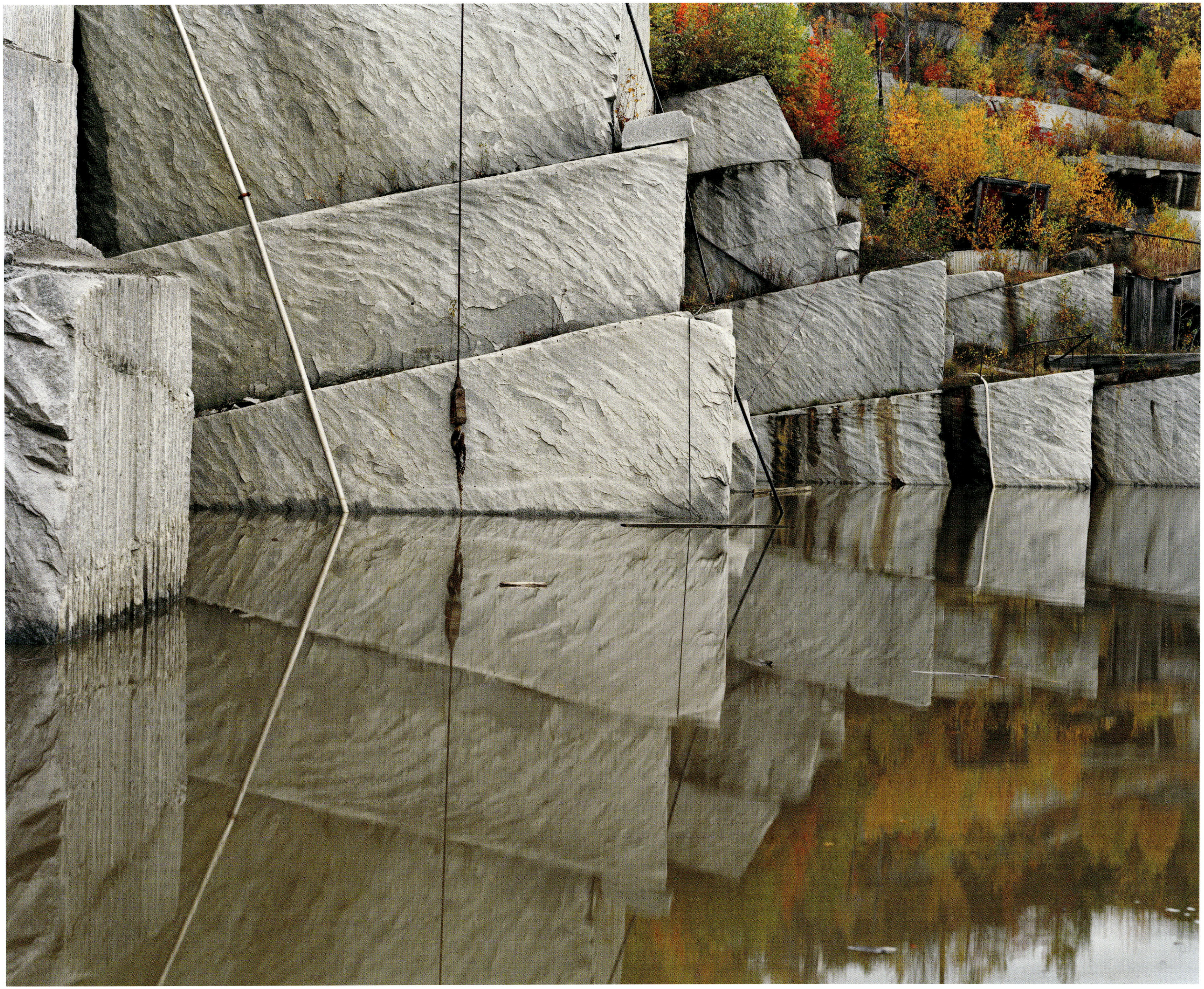

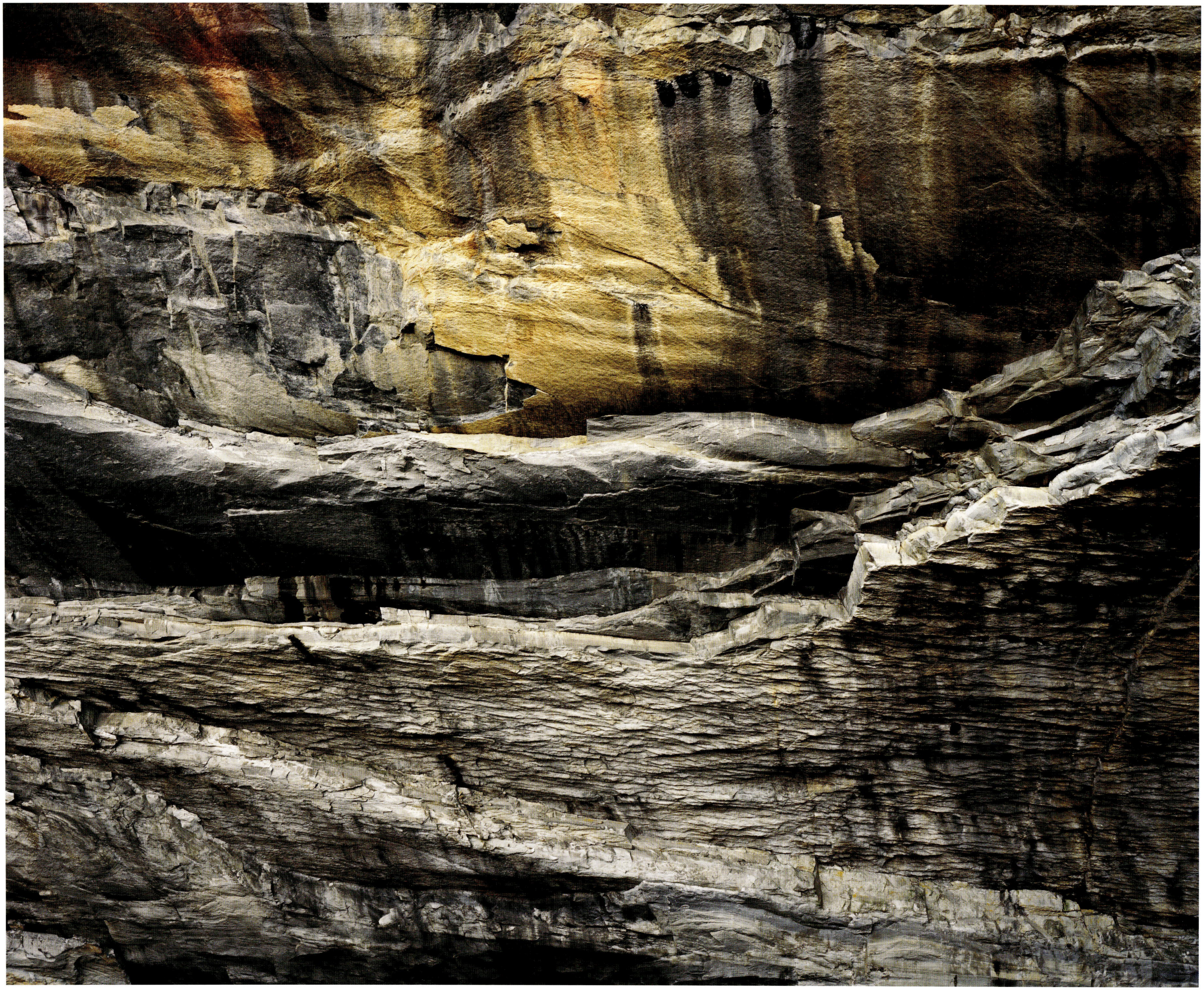

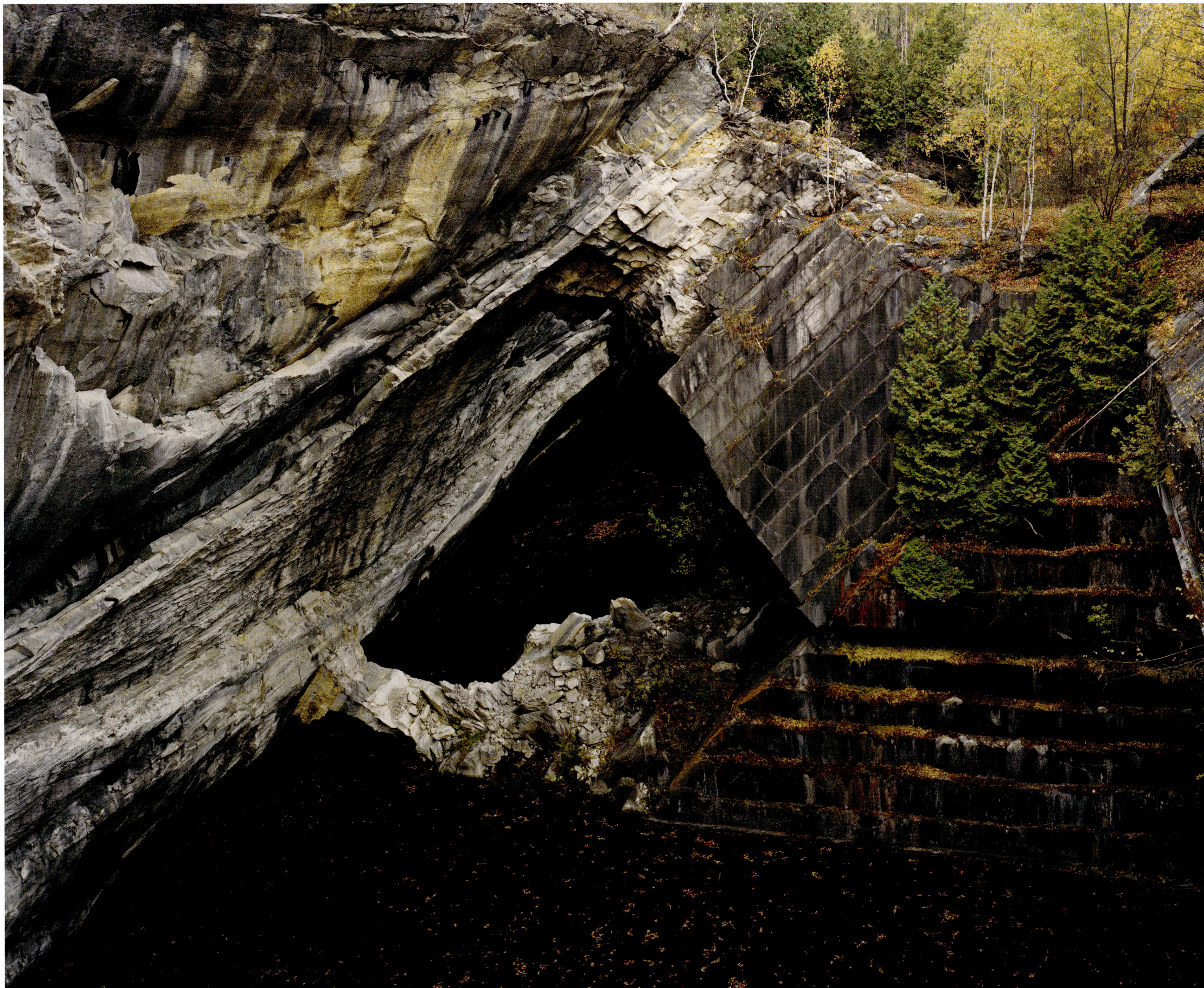

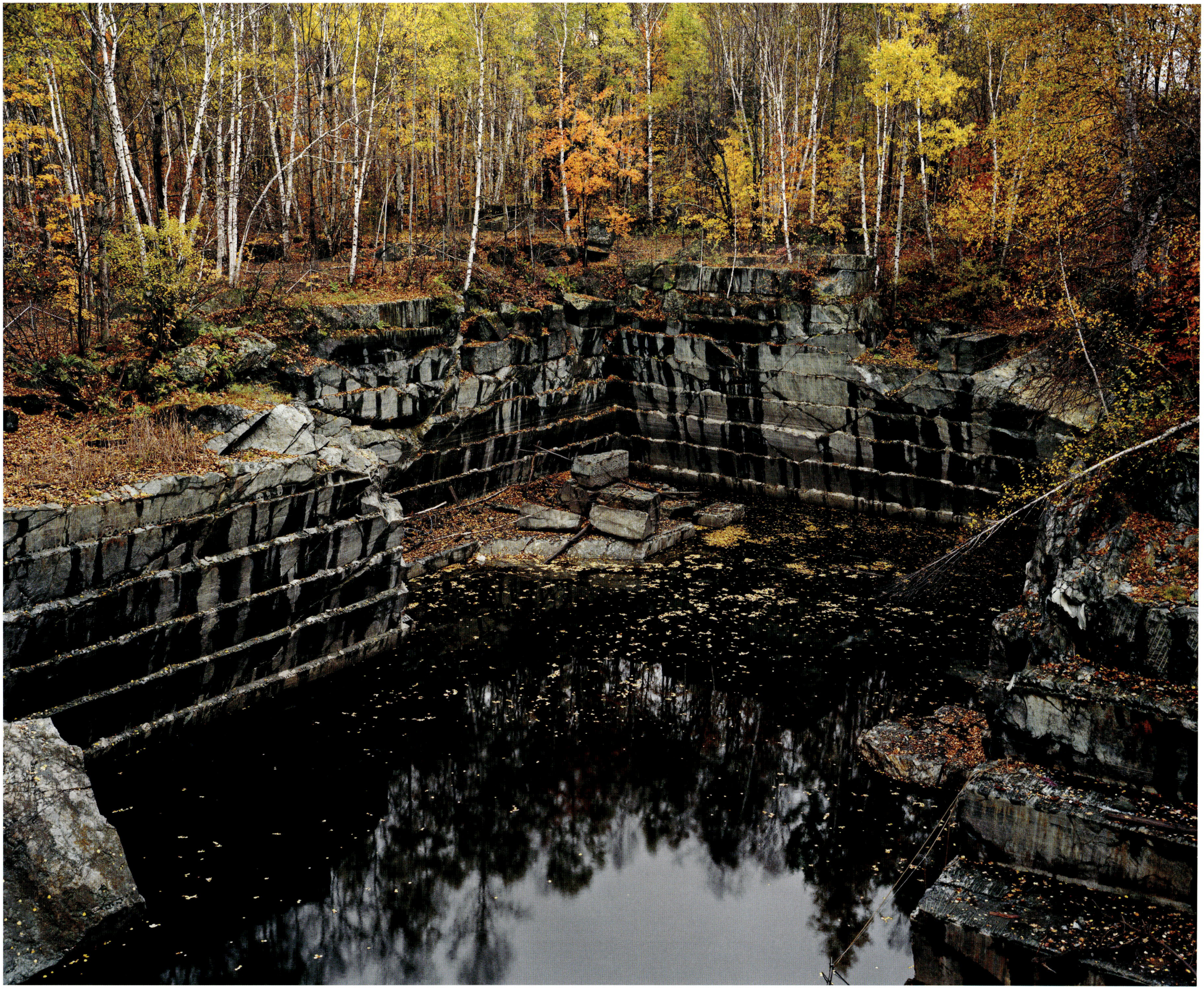

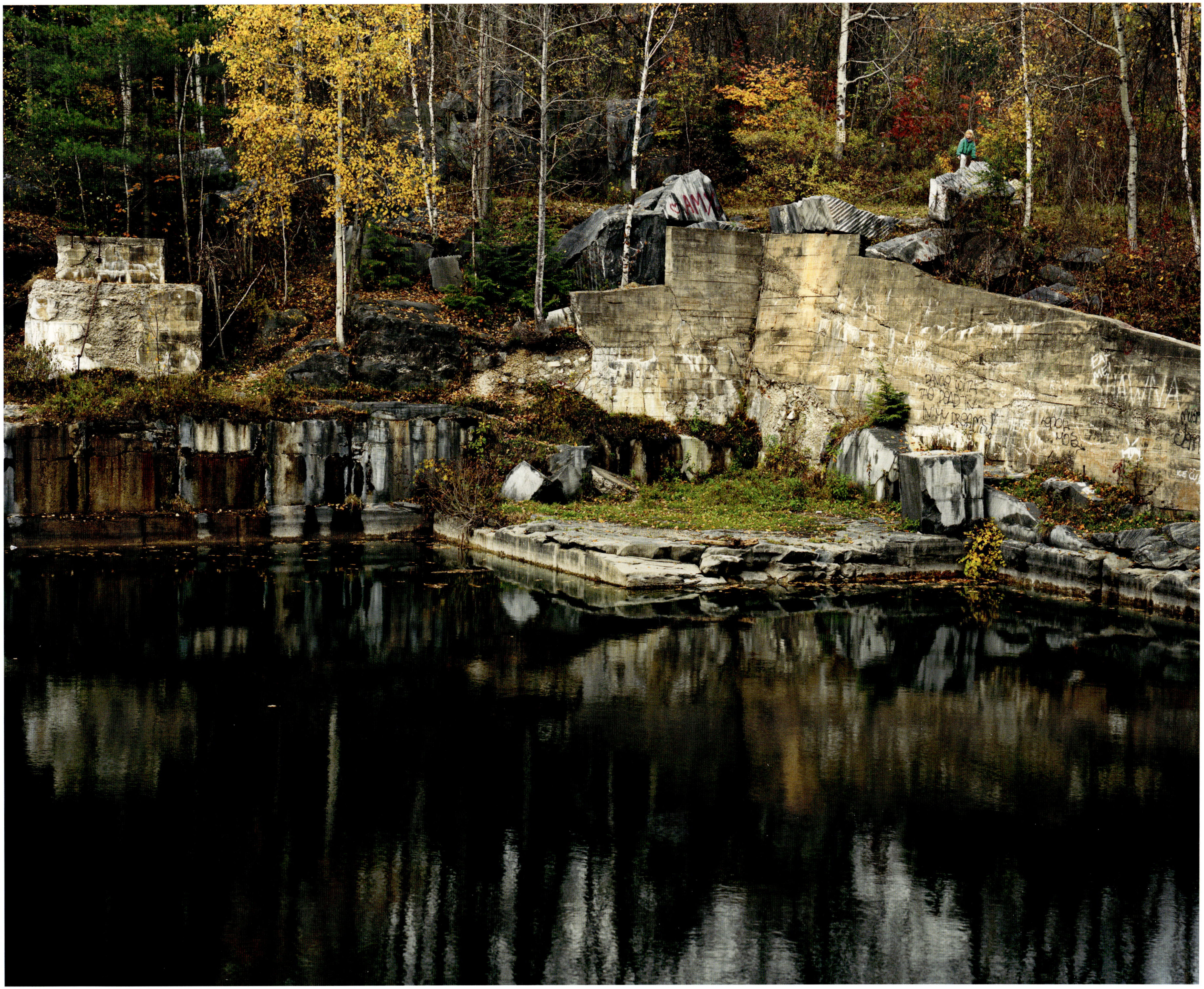

CARRARA

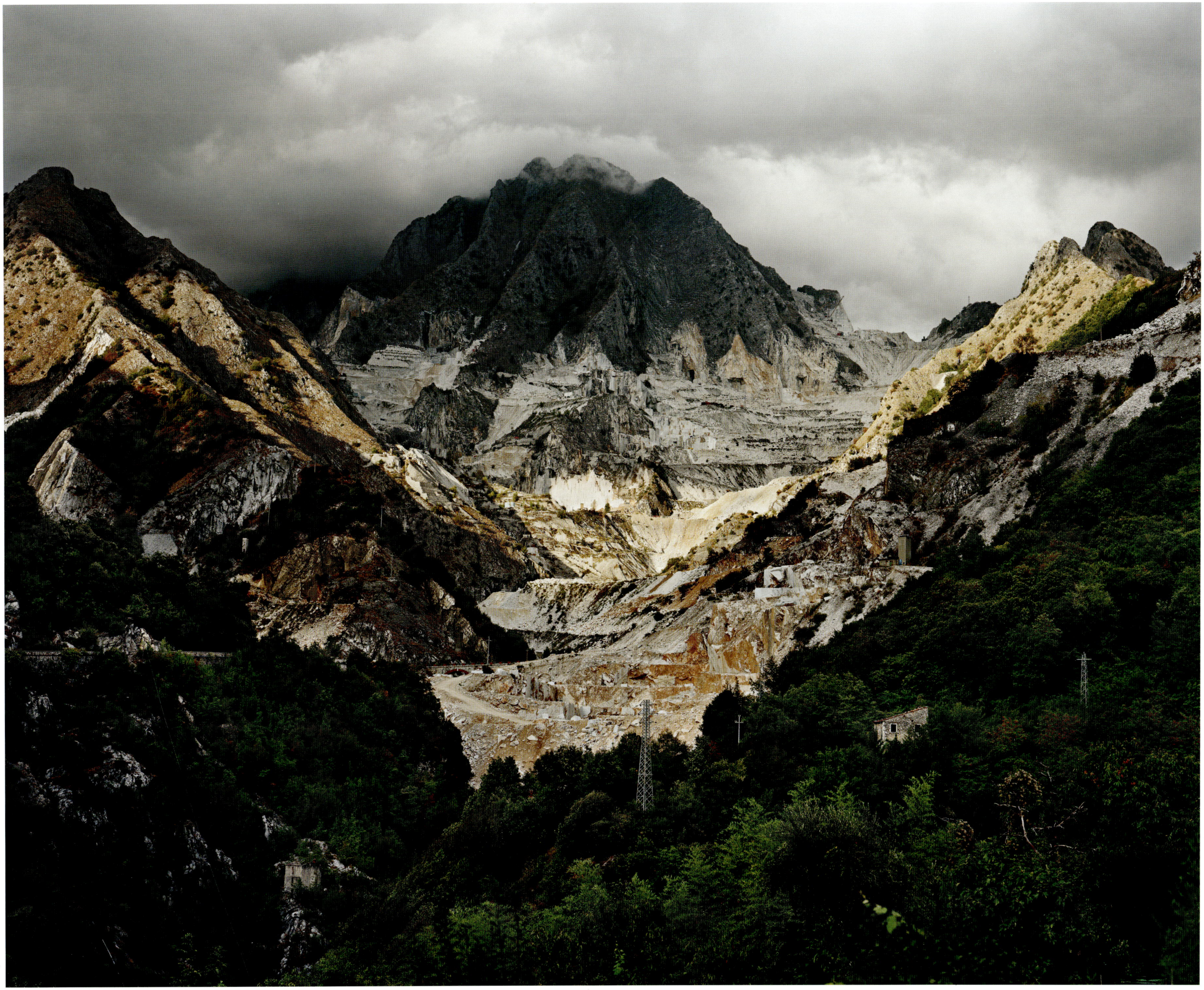

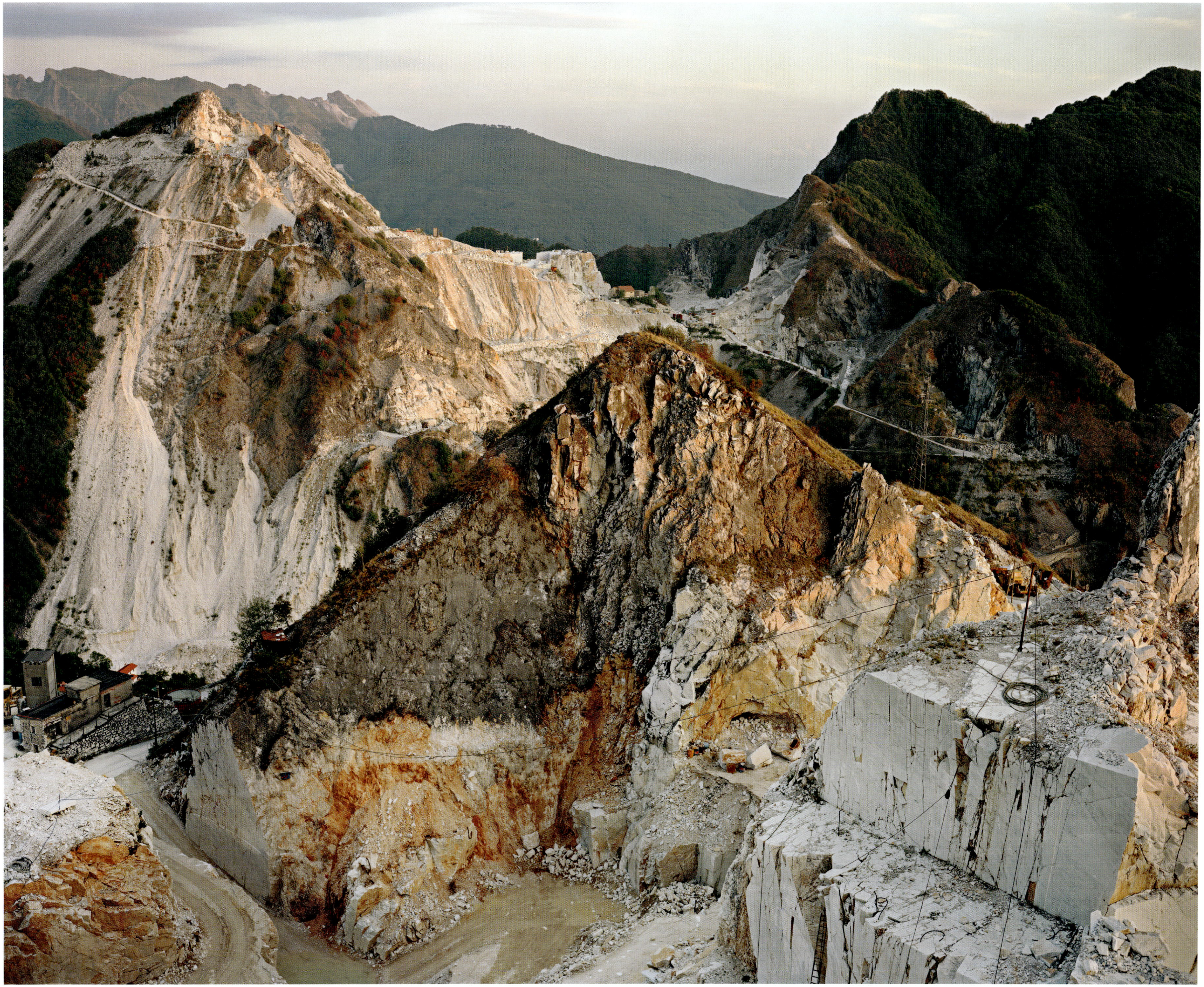

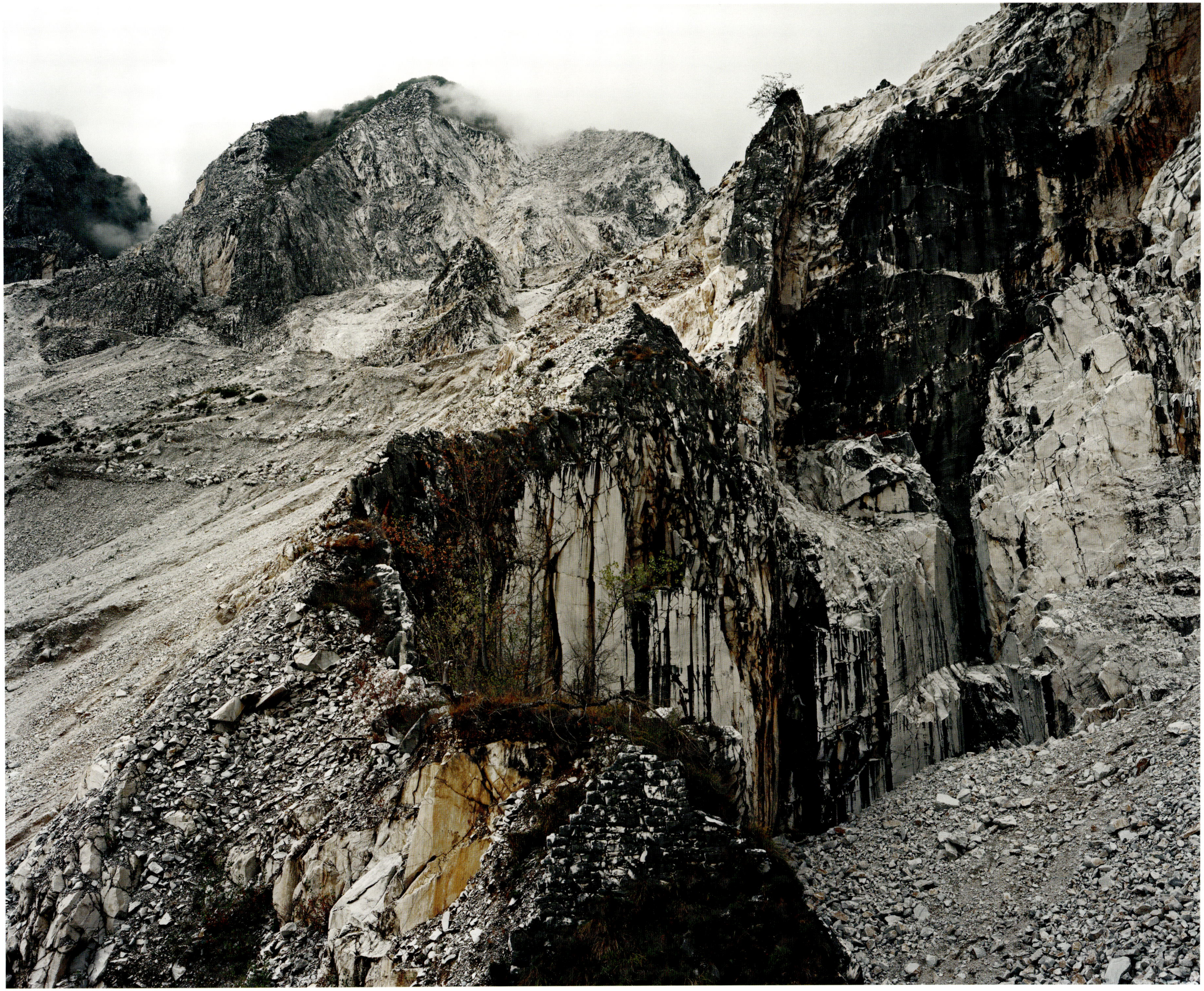

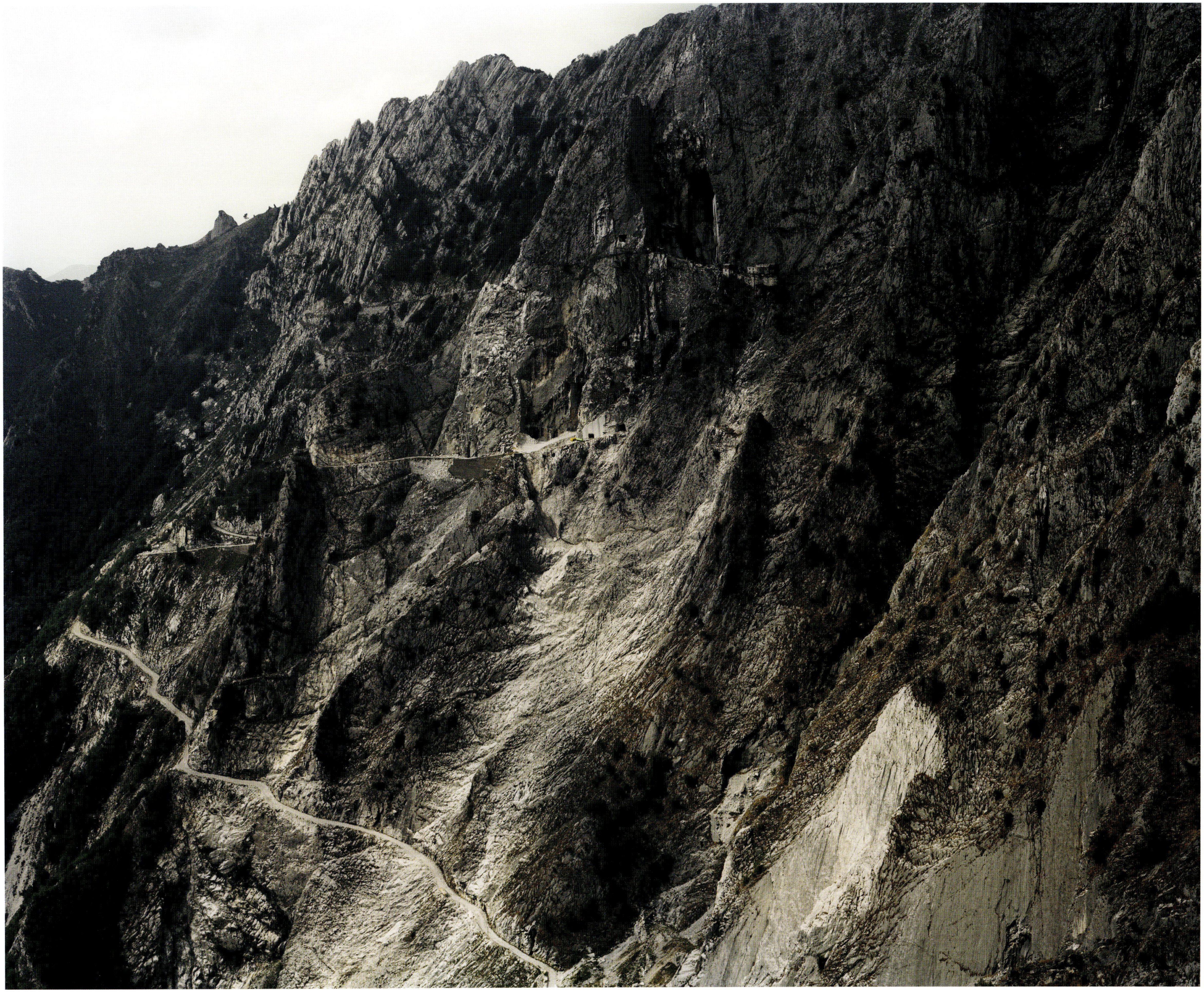

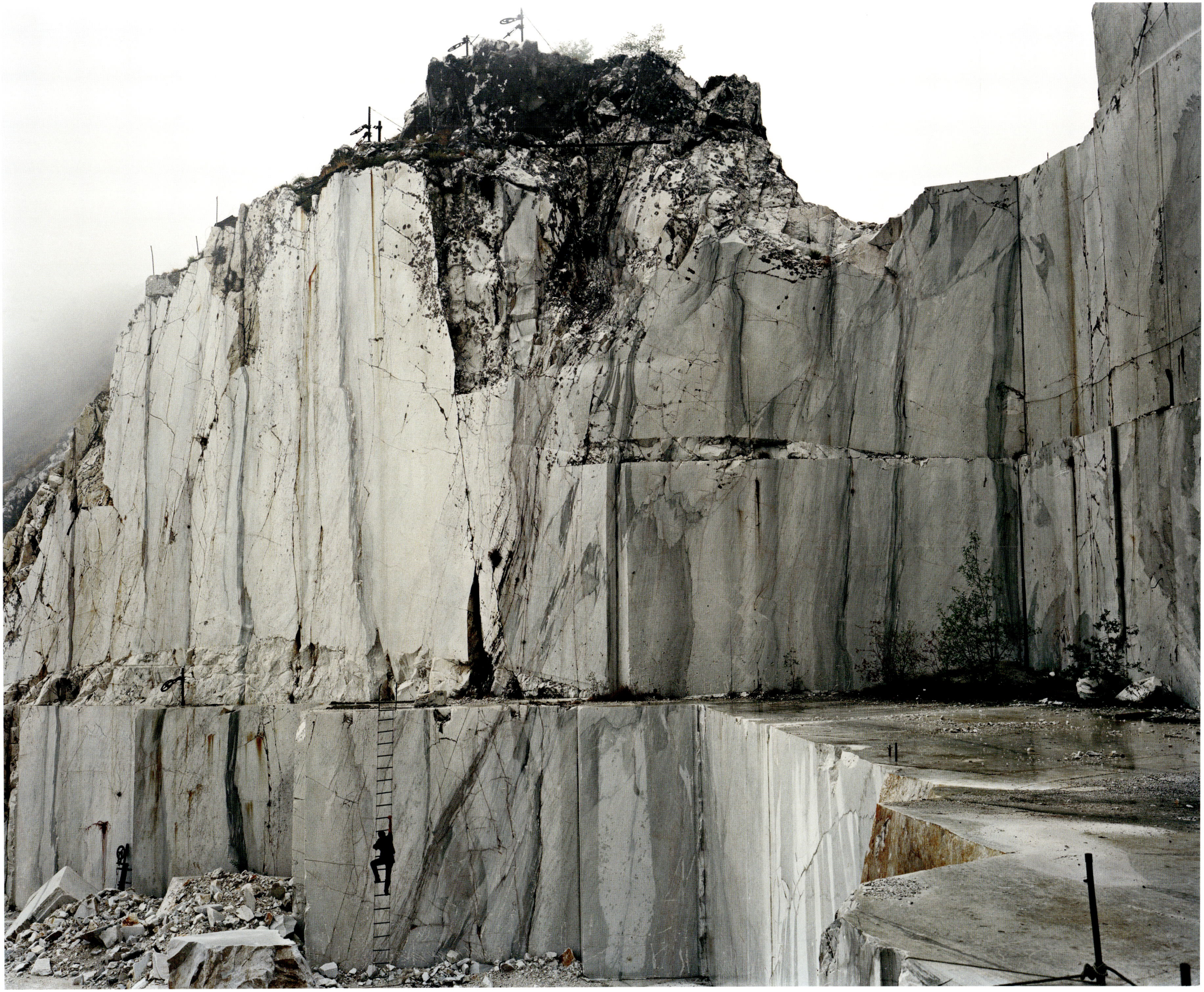

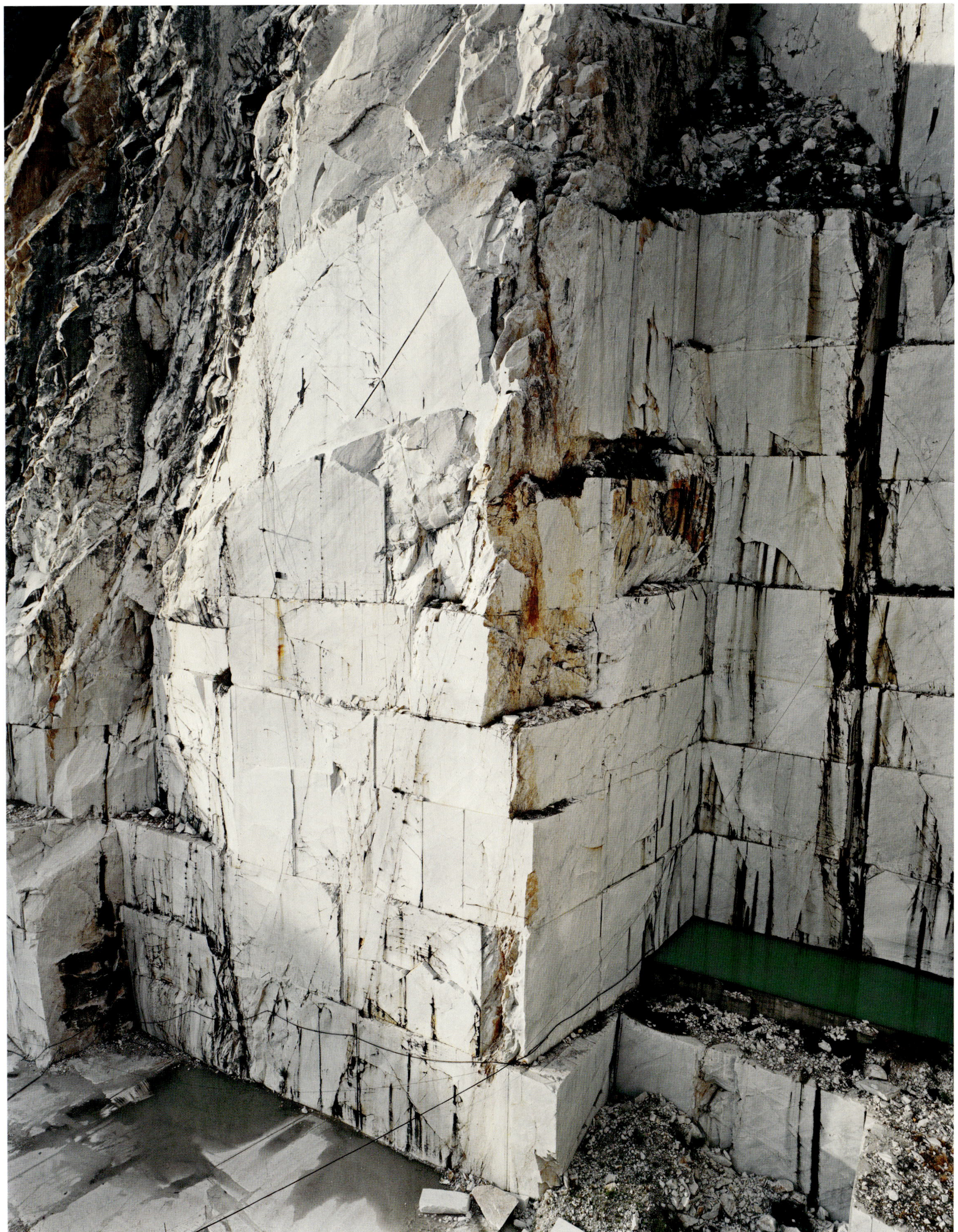

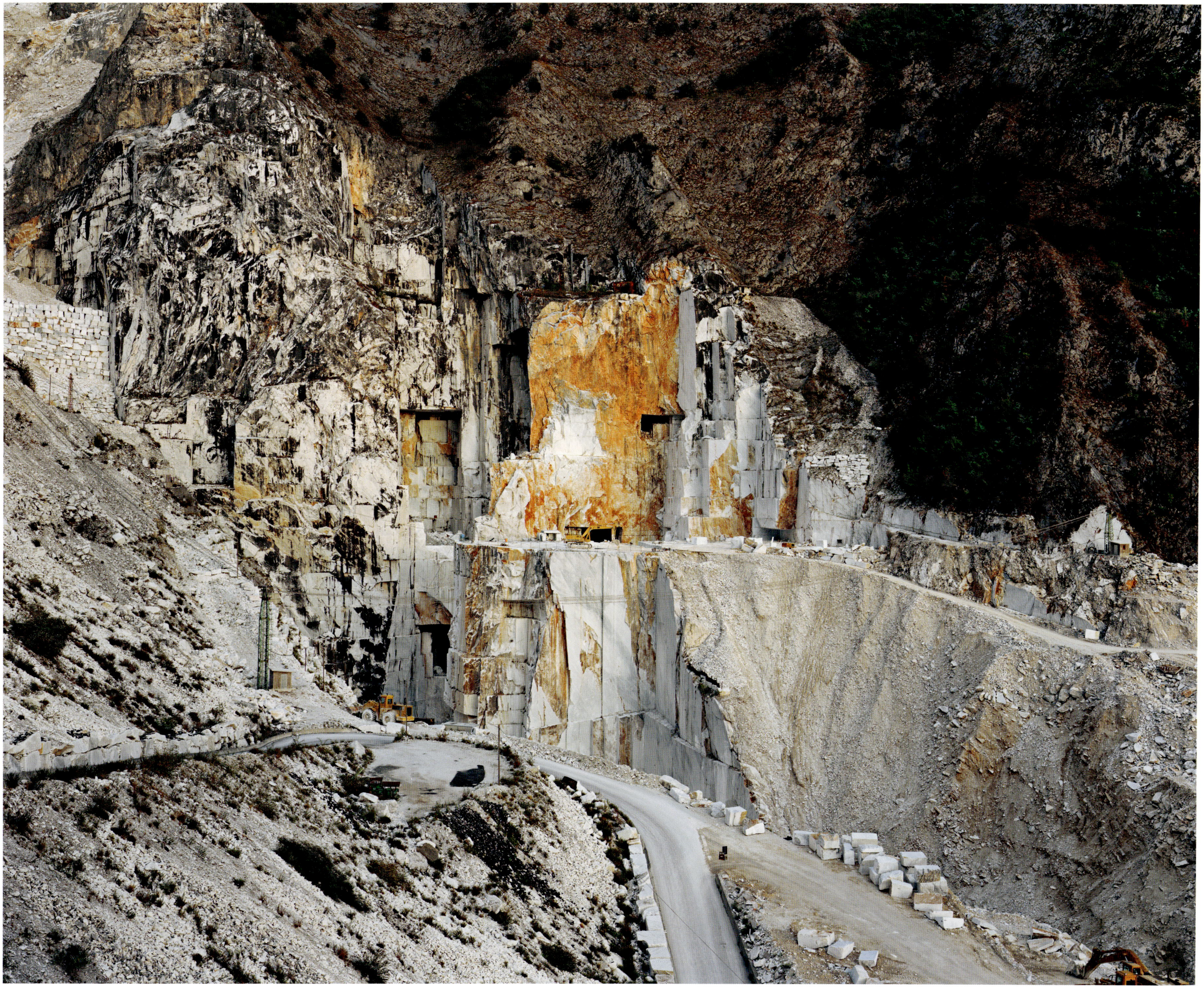

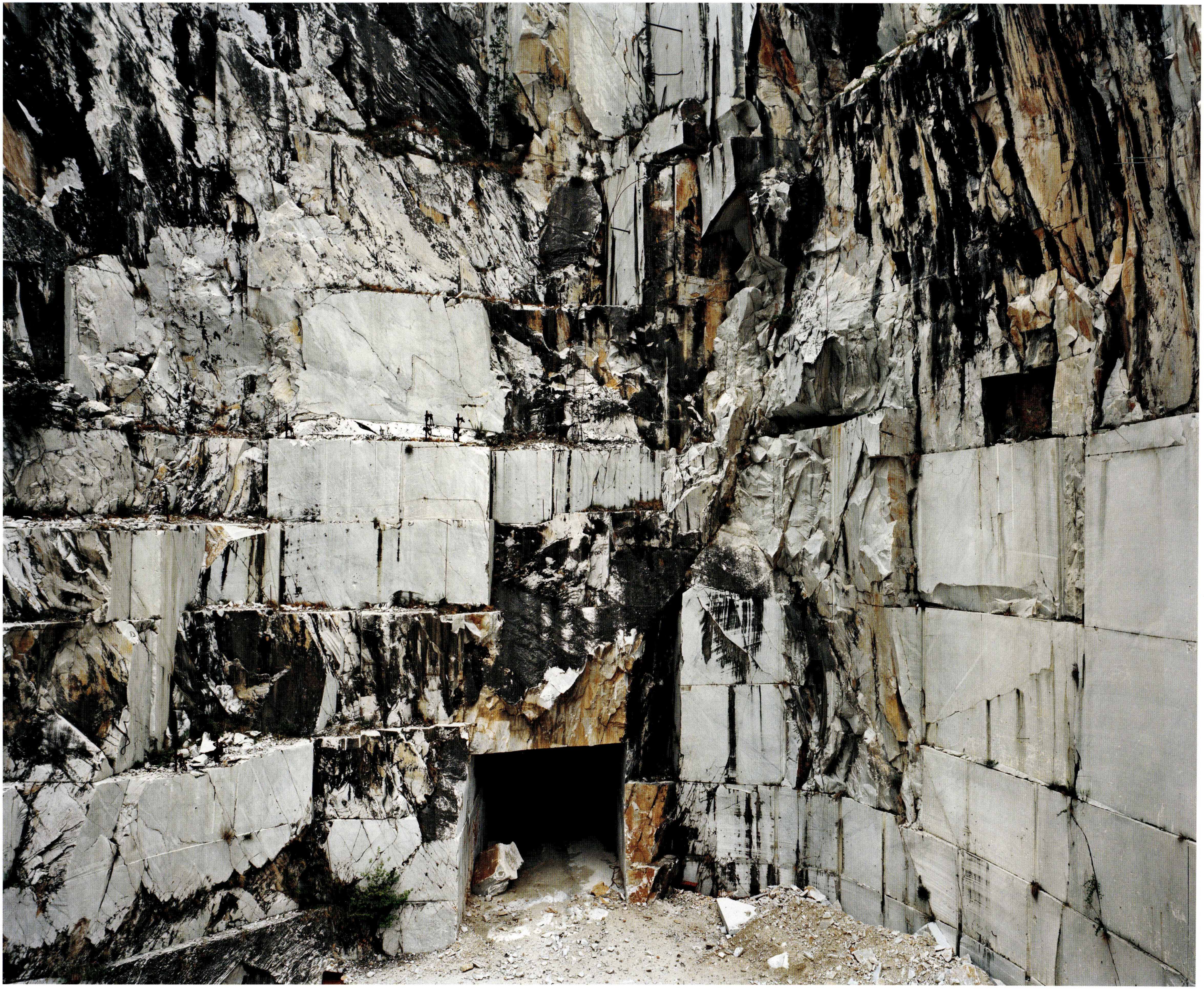

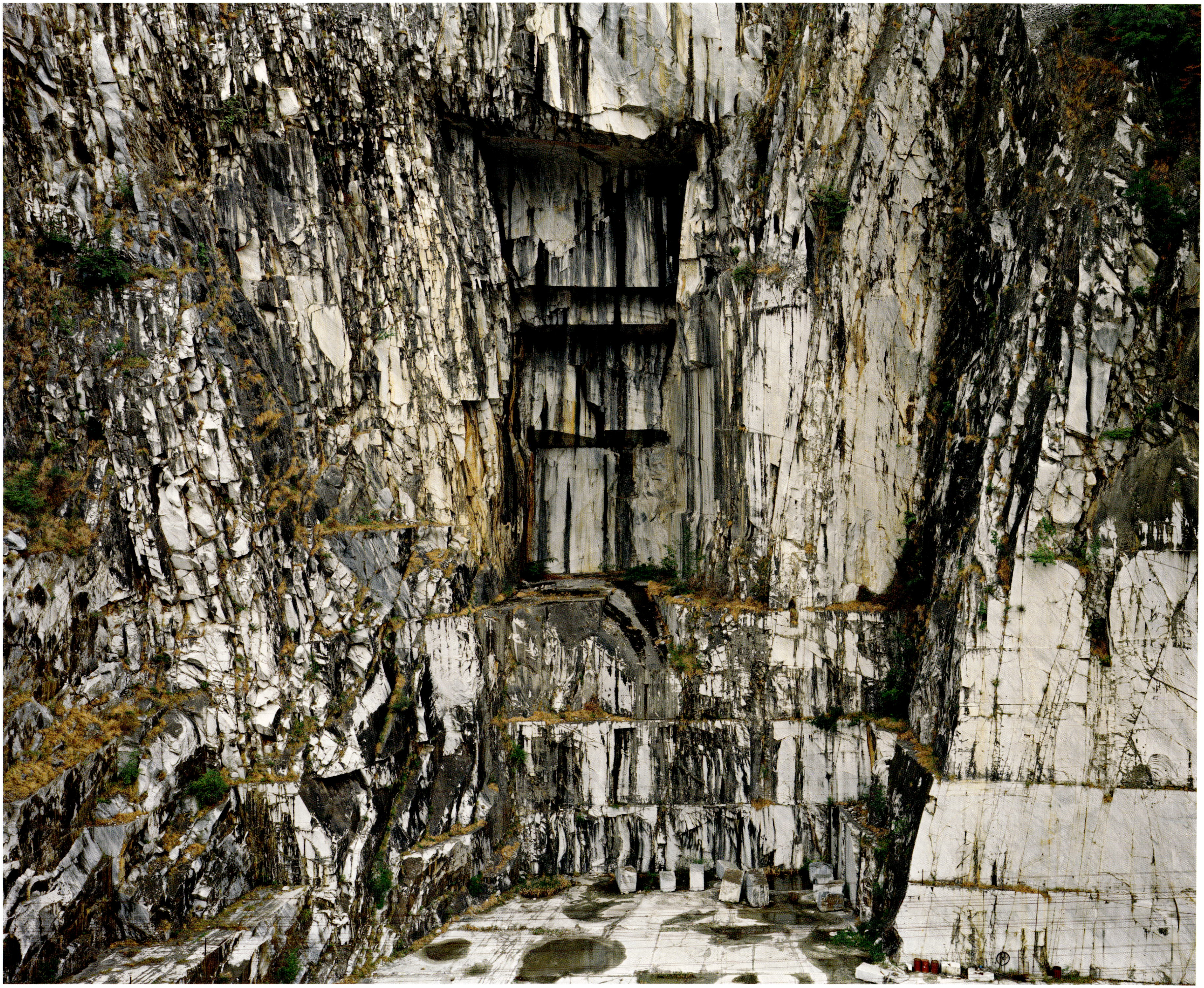

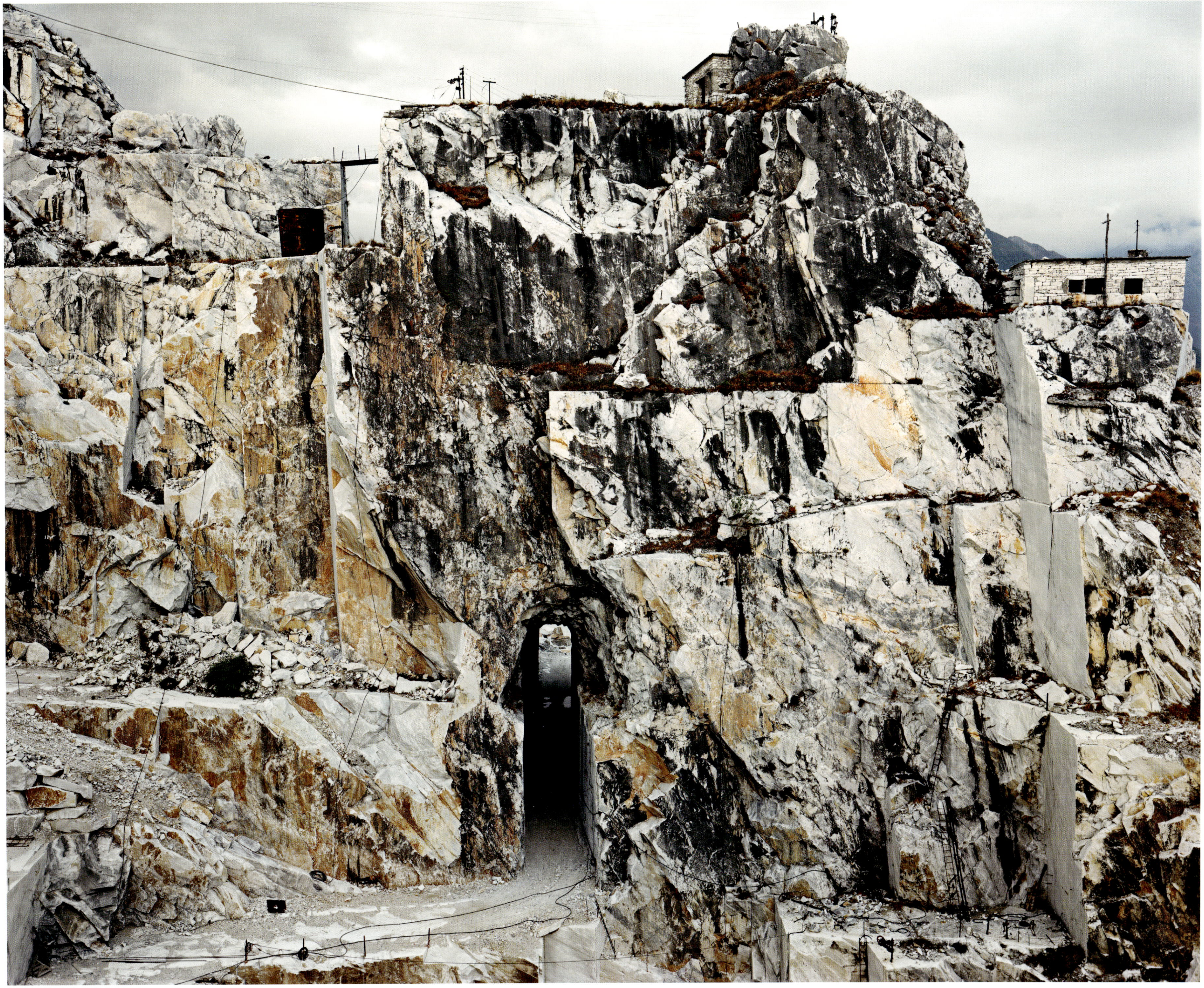

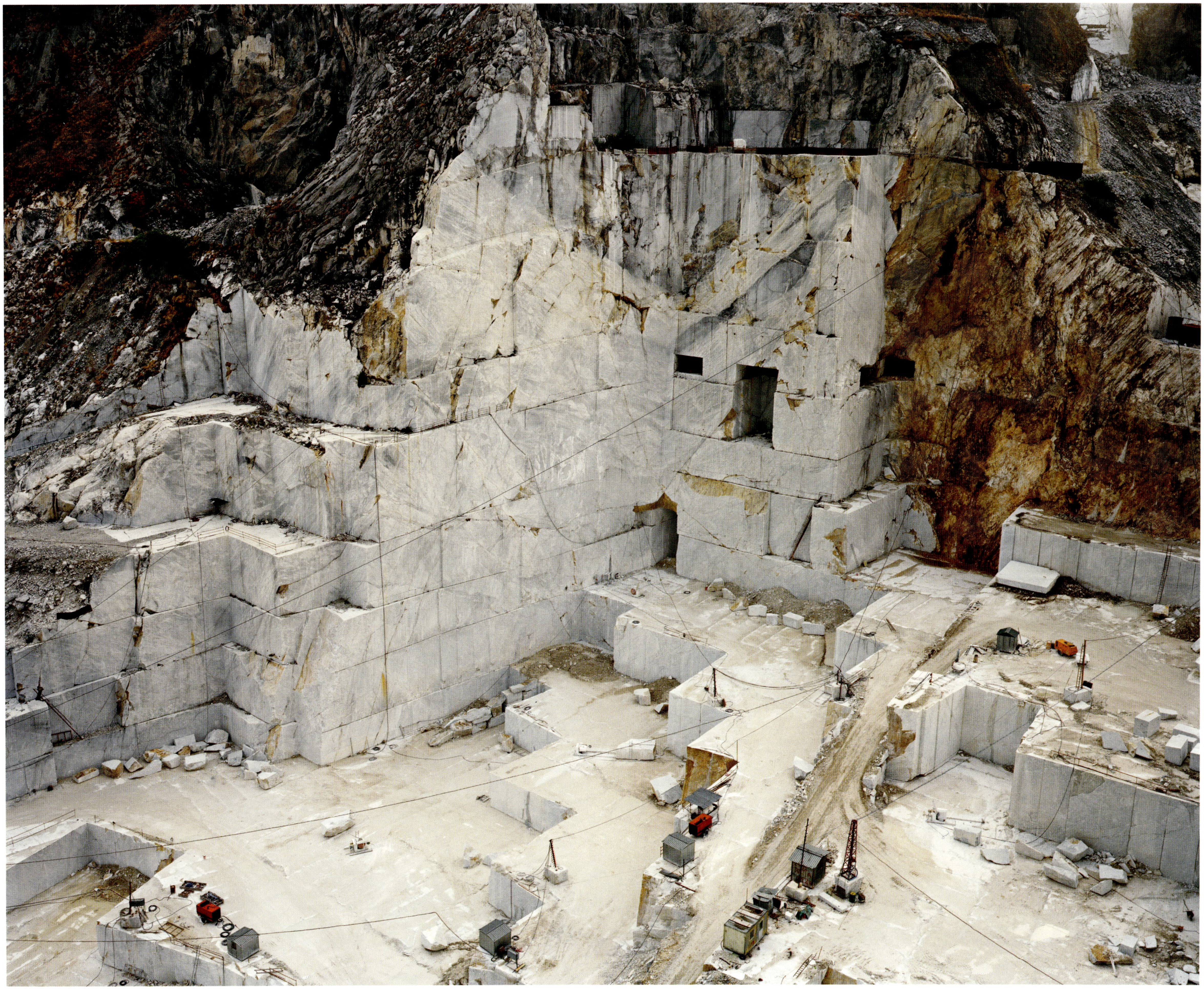

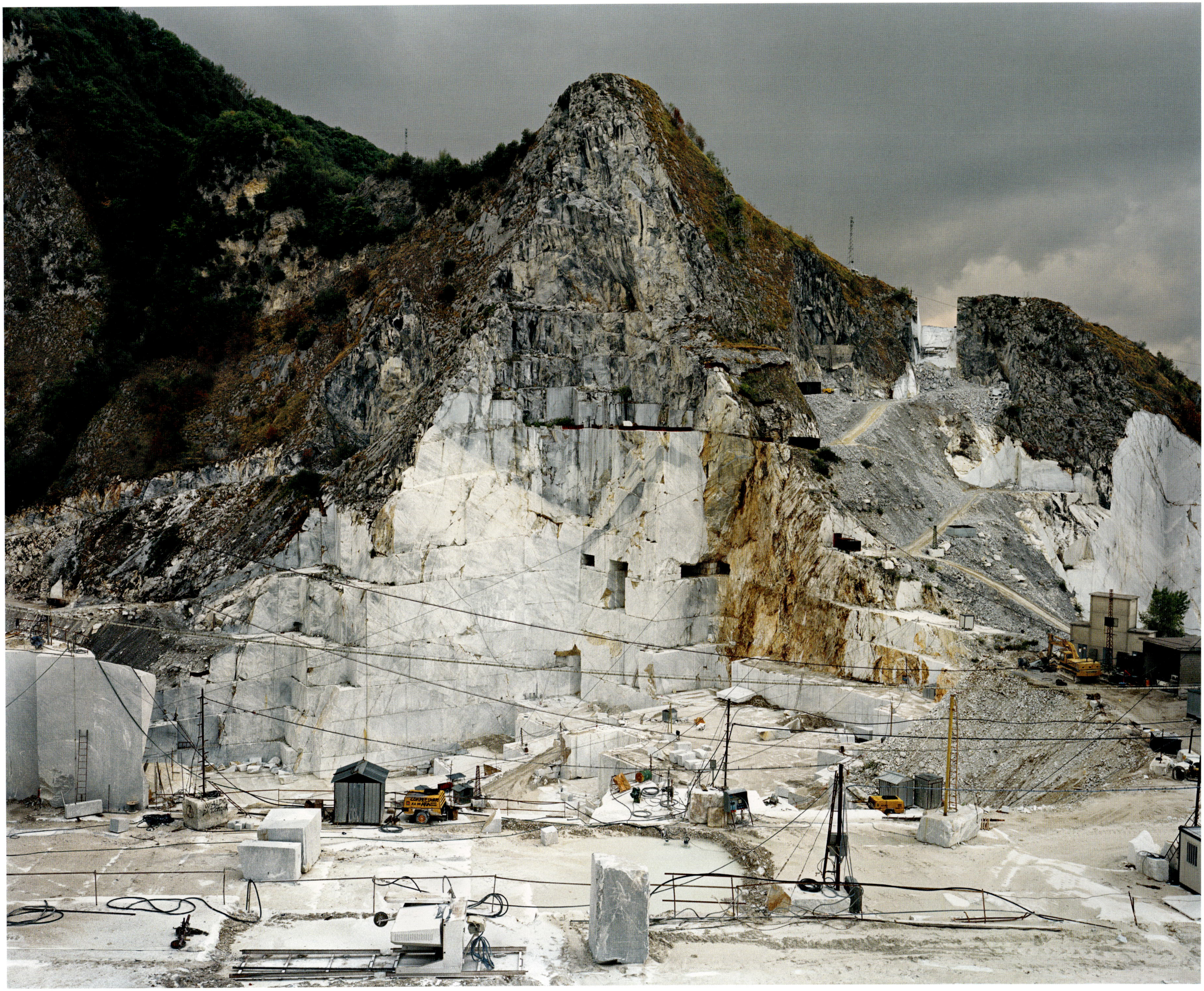

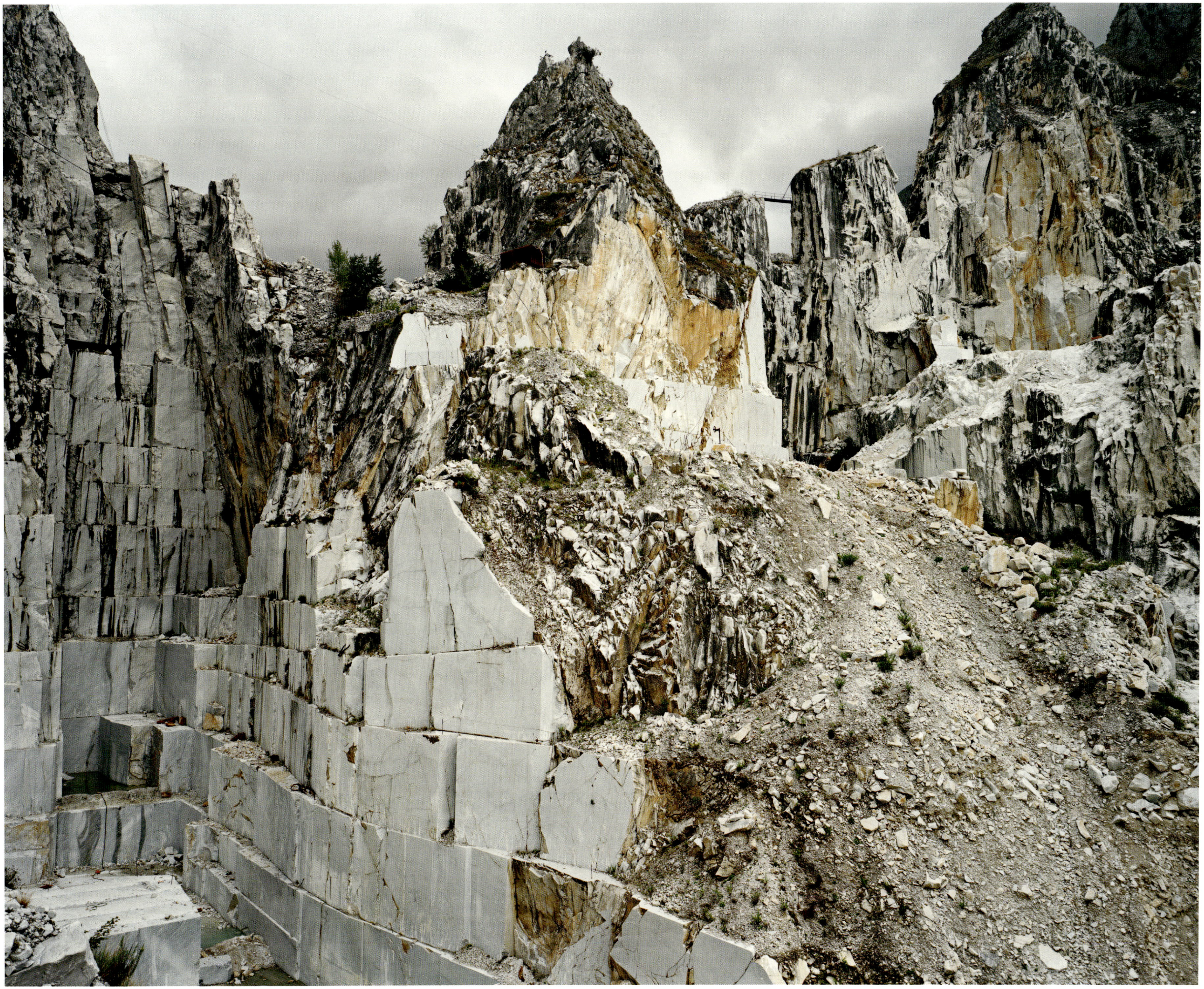

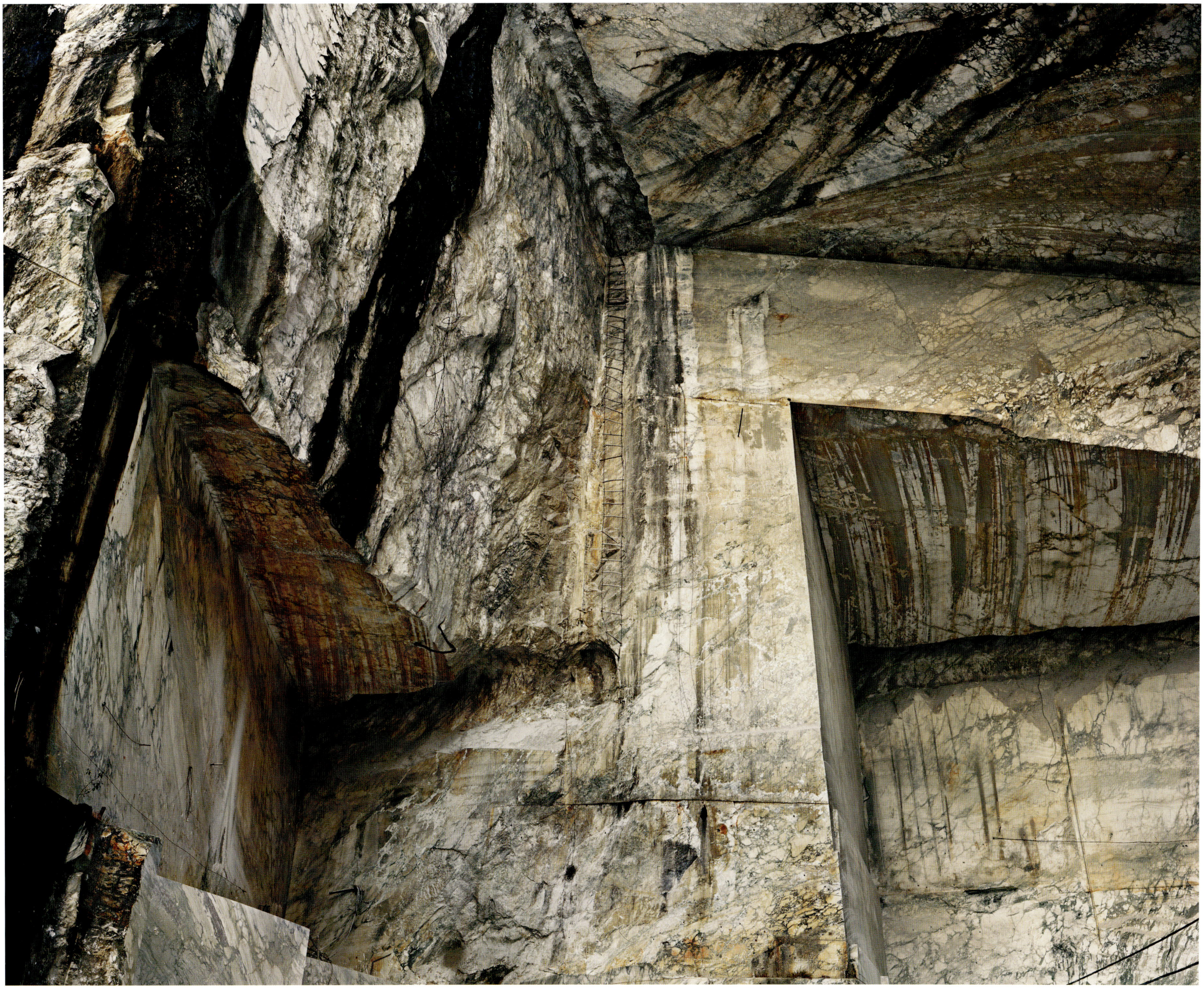

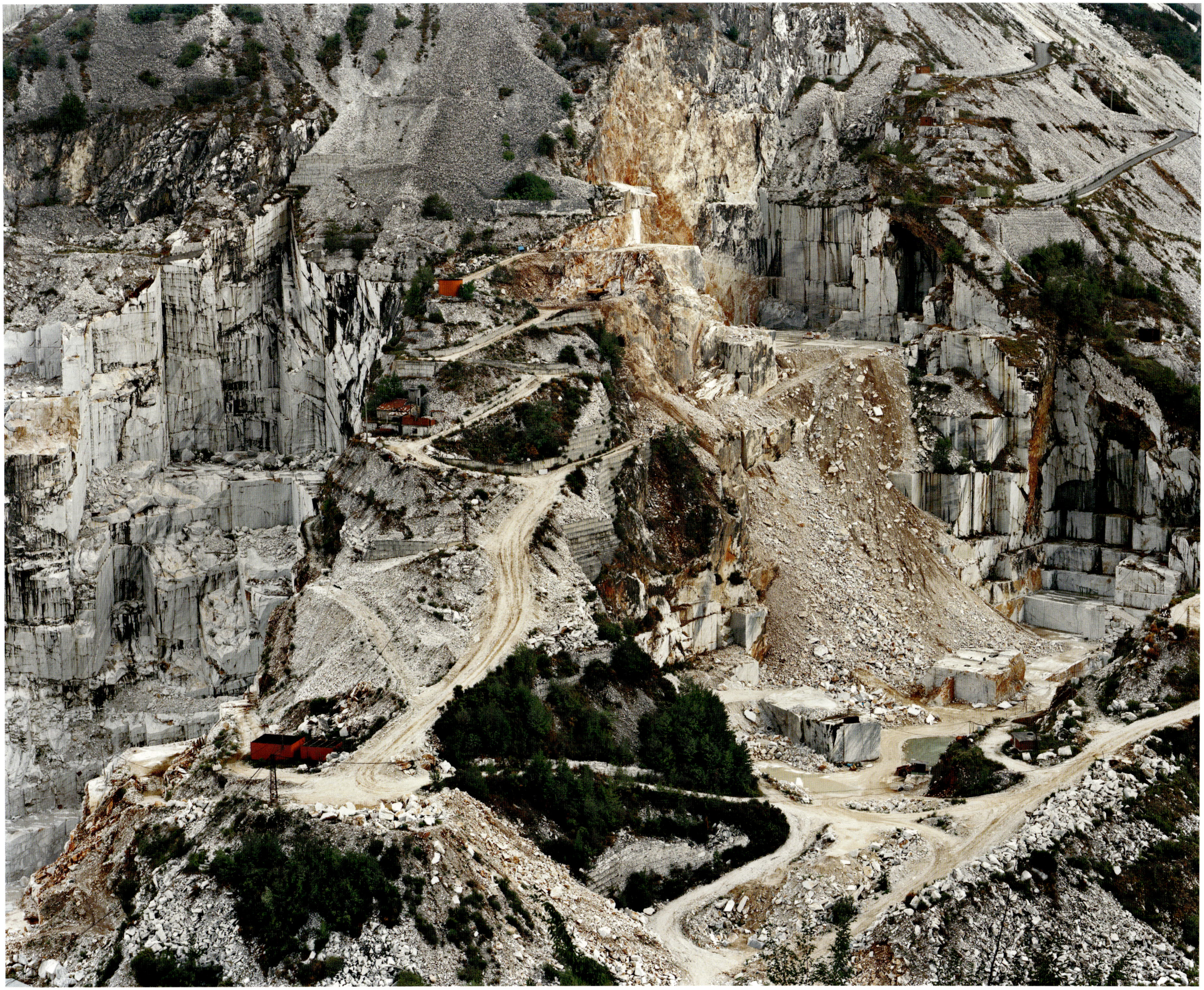

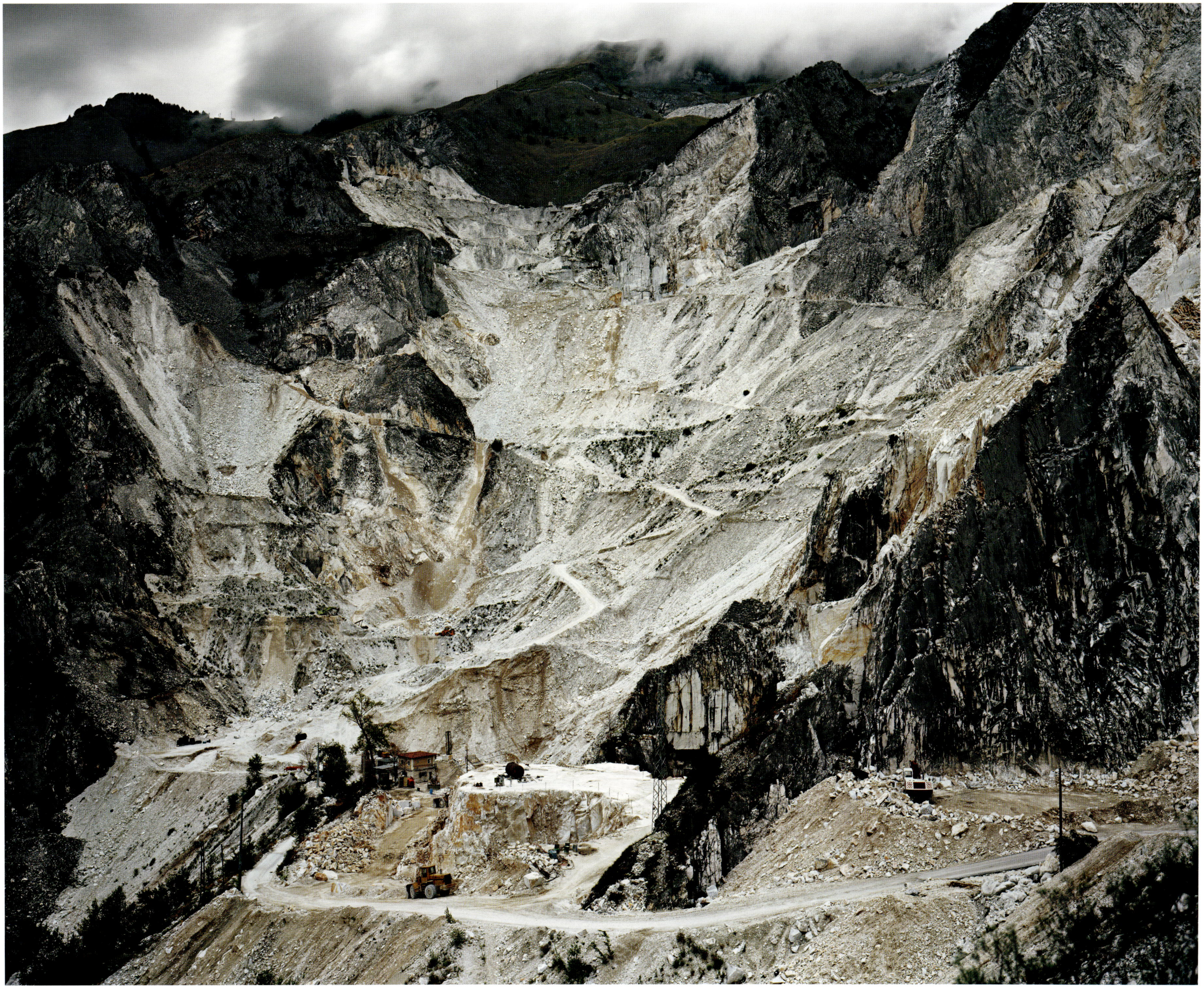

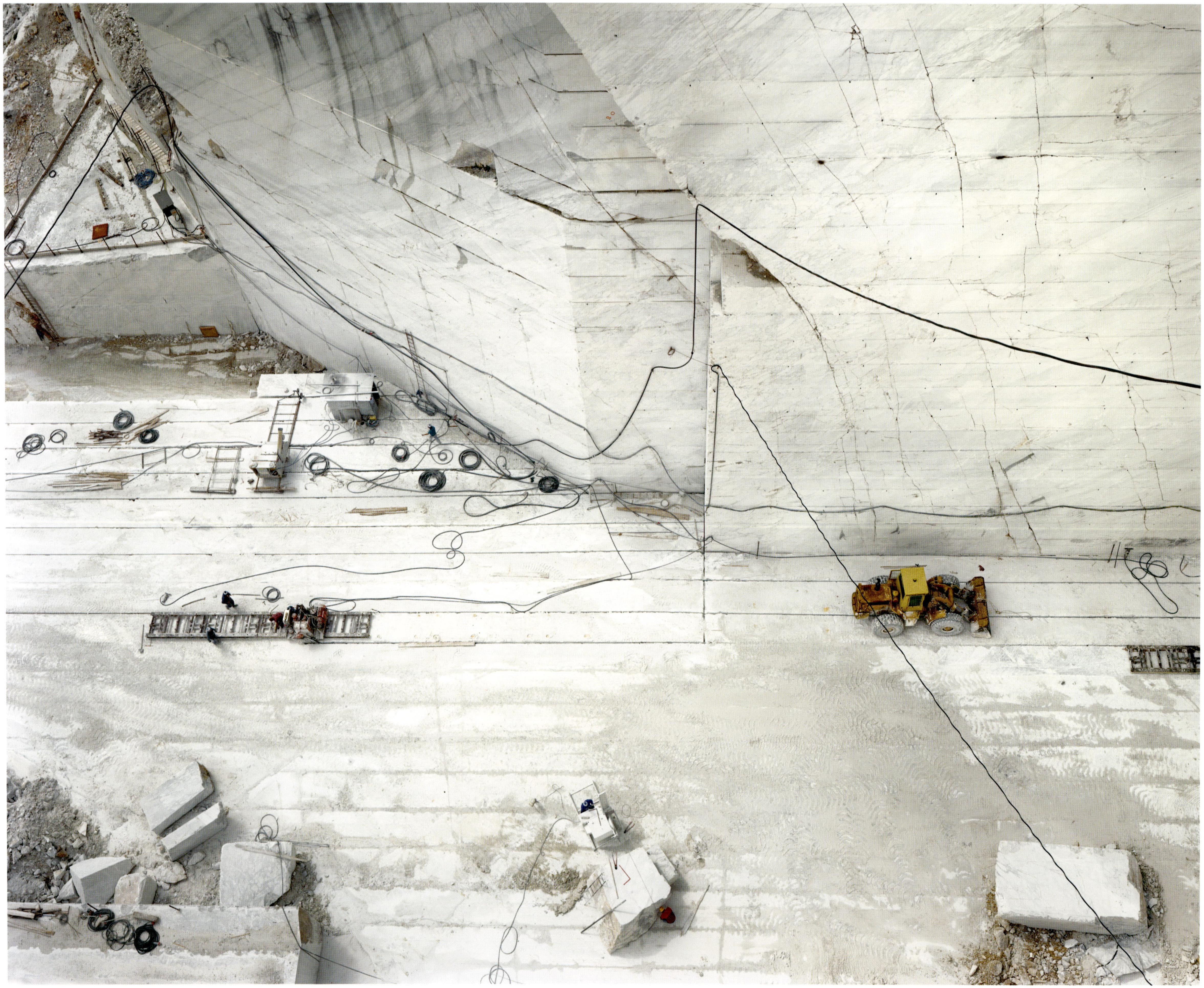

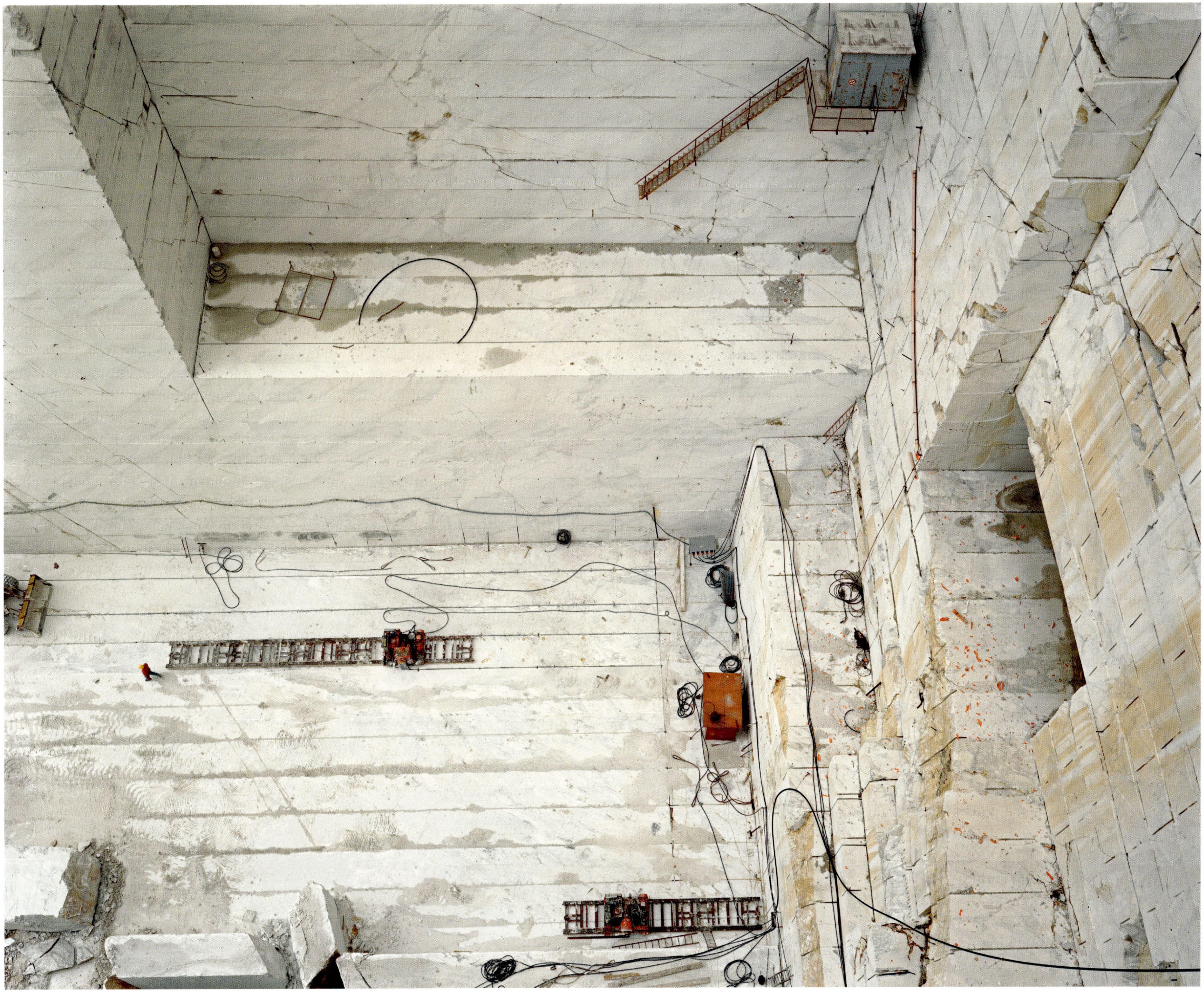

MAKRANA

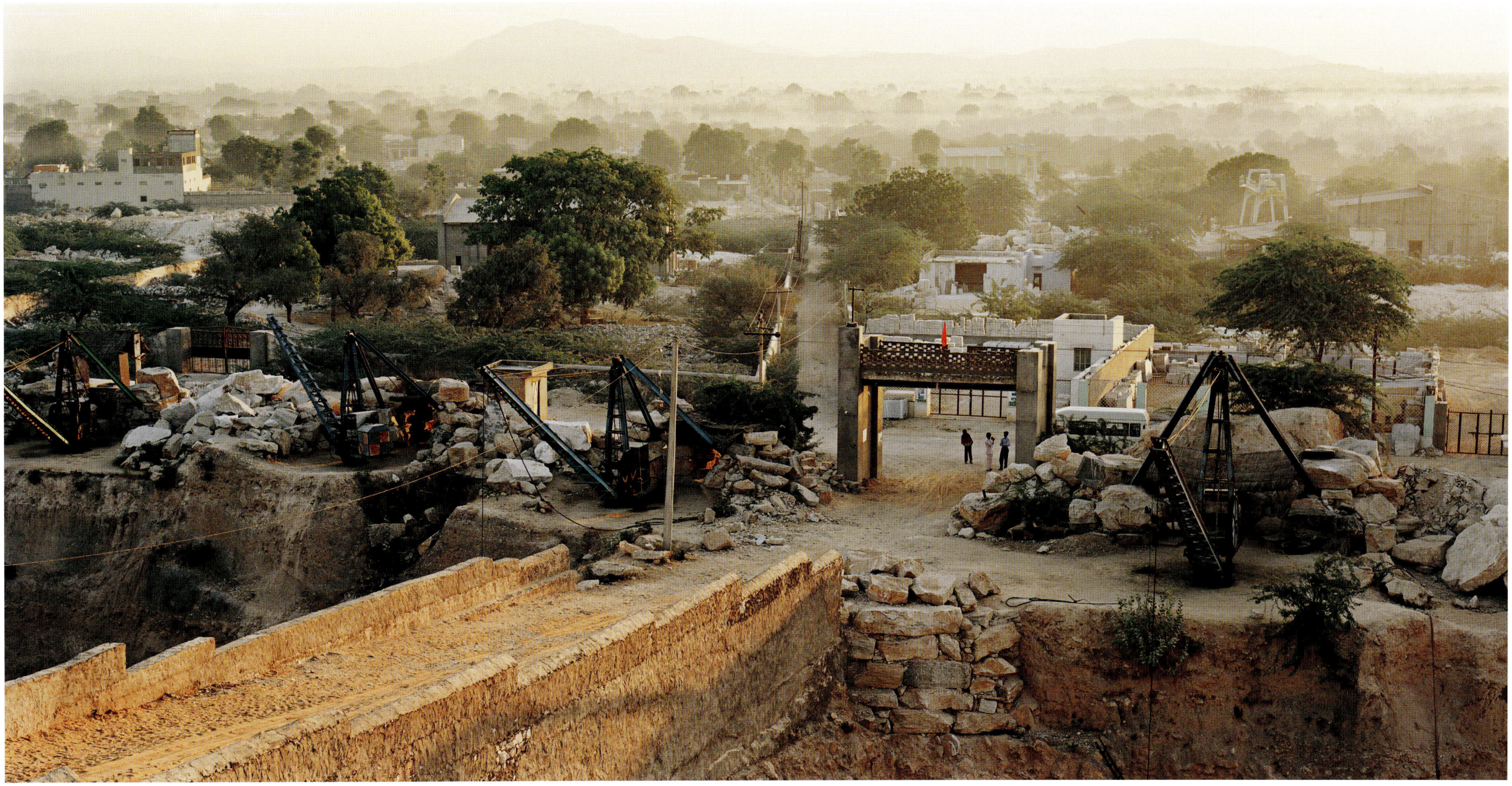

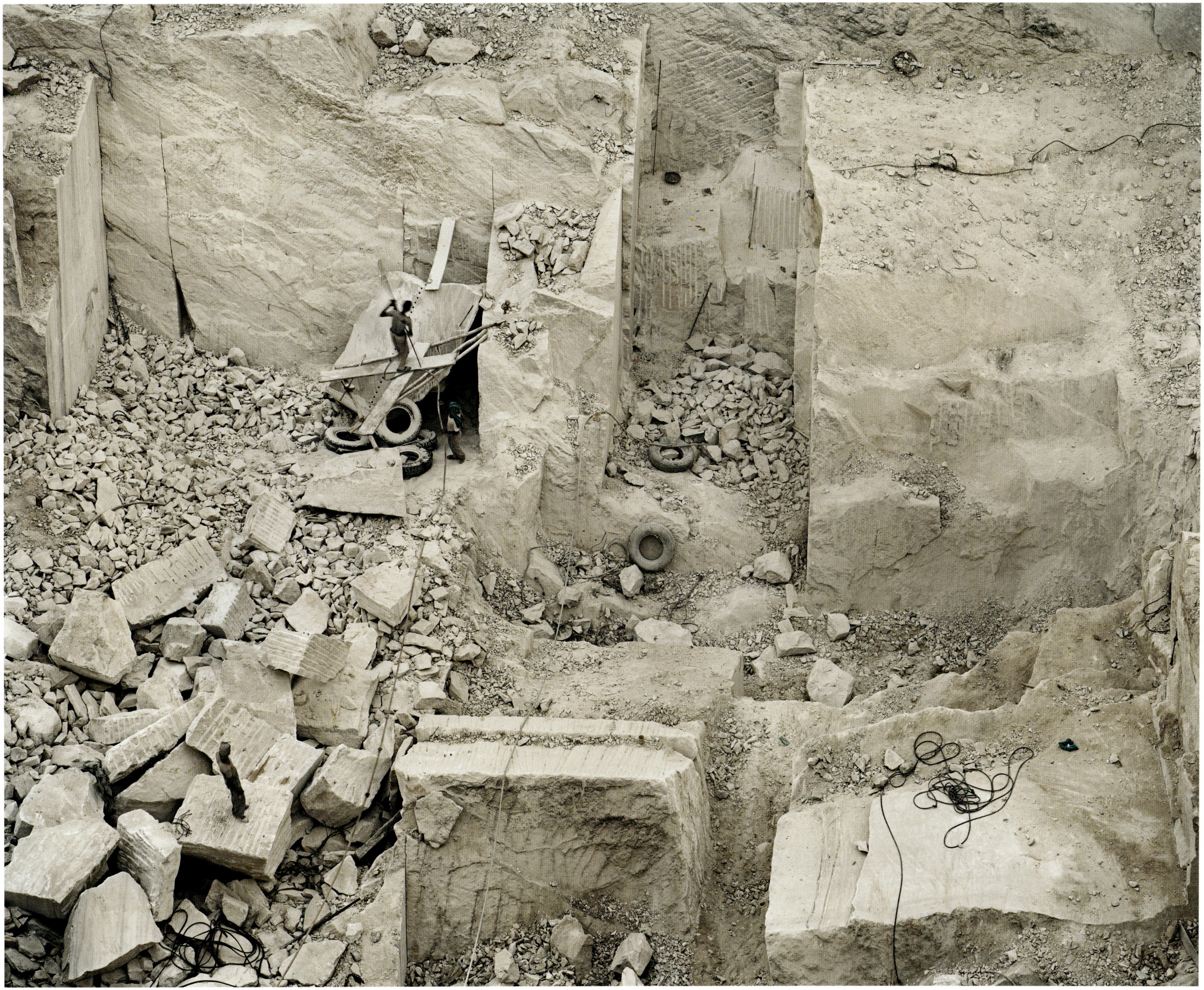

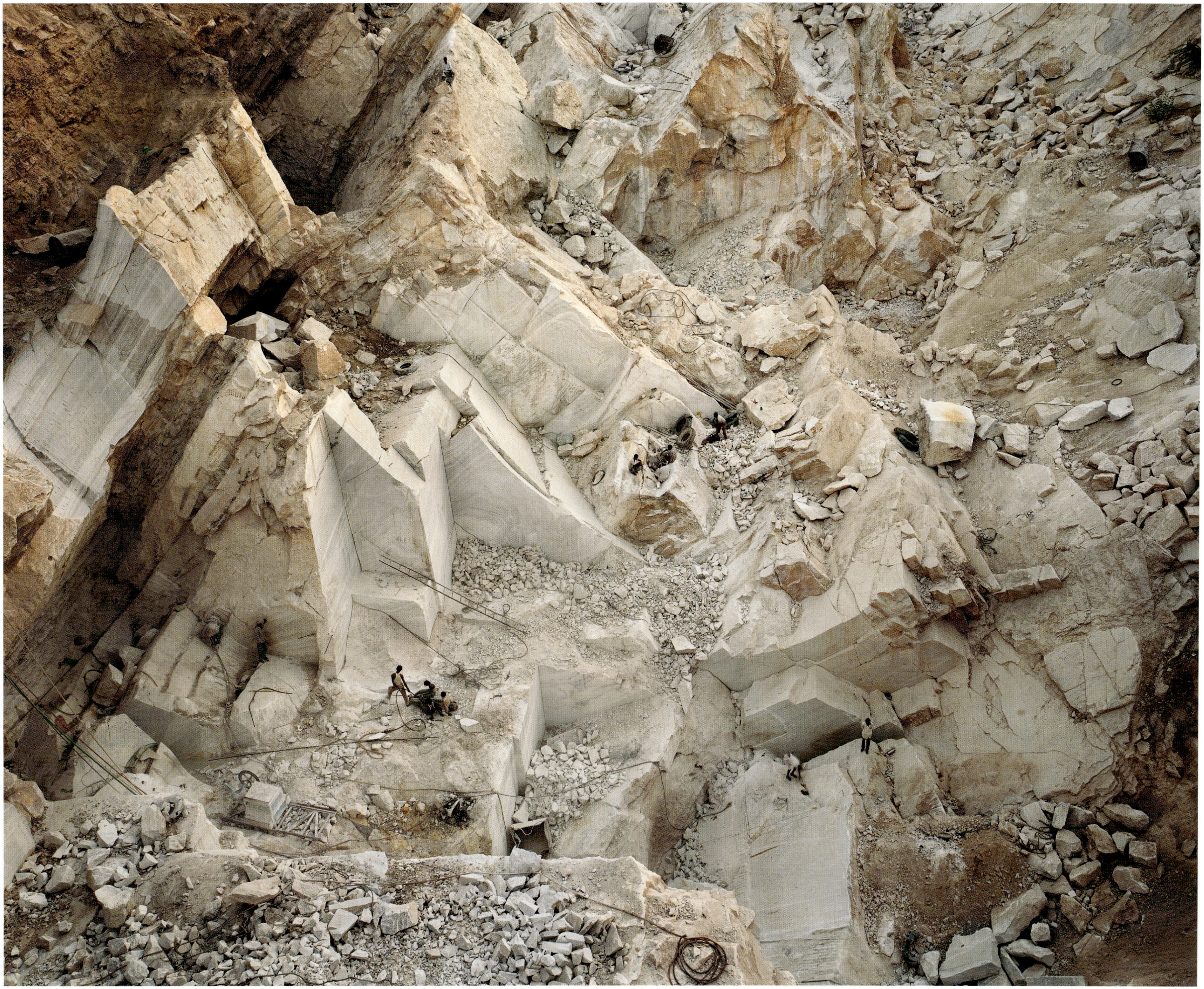

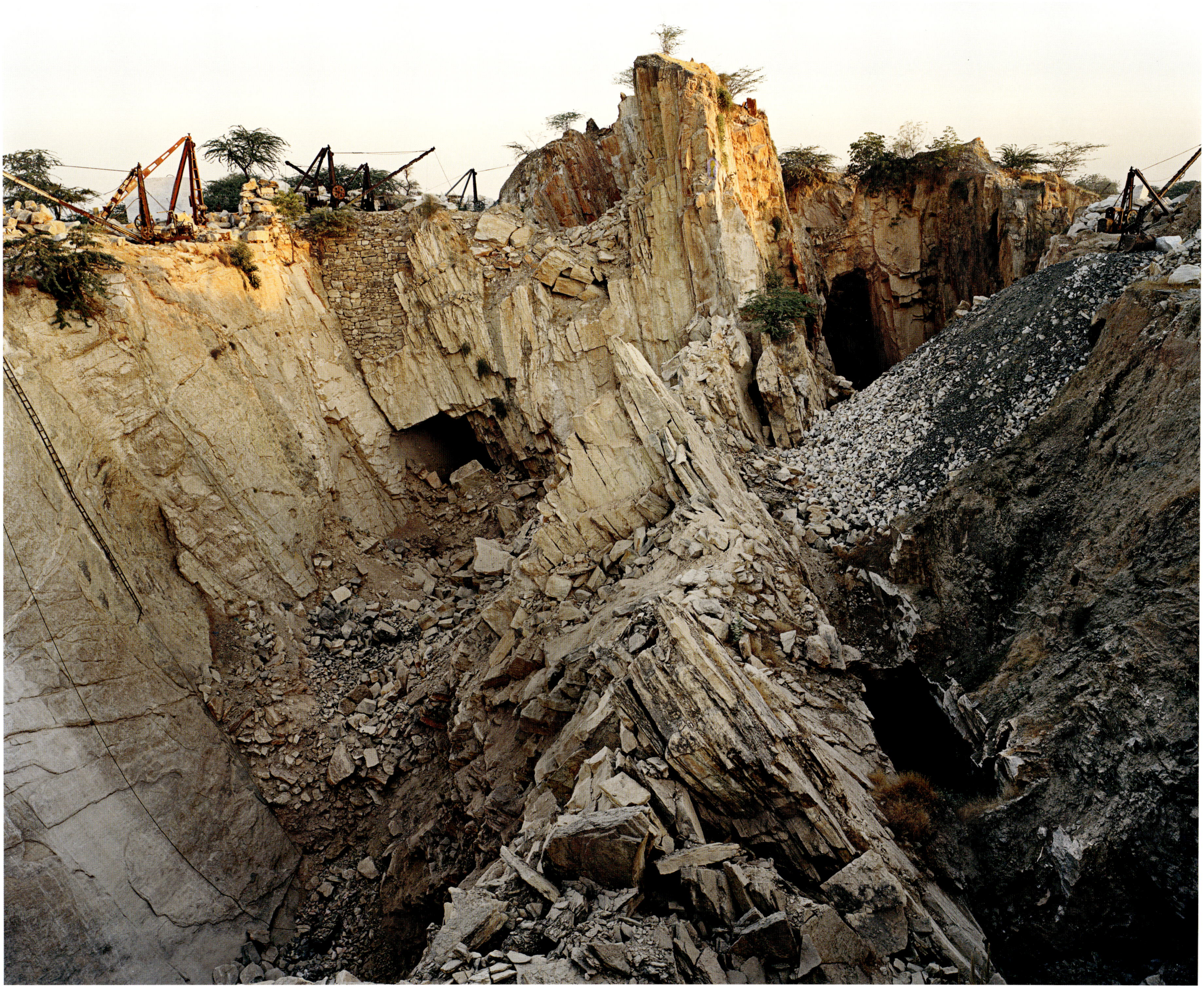

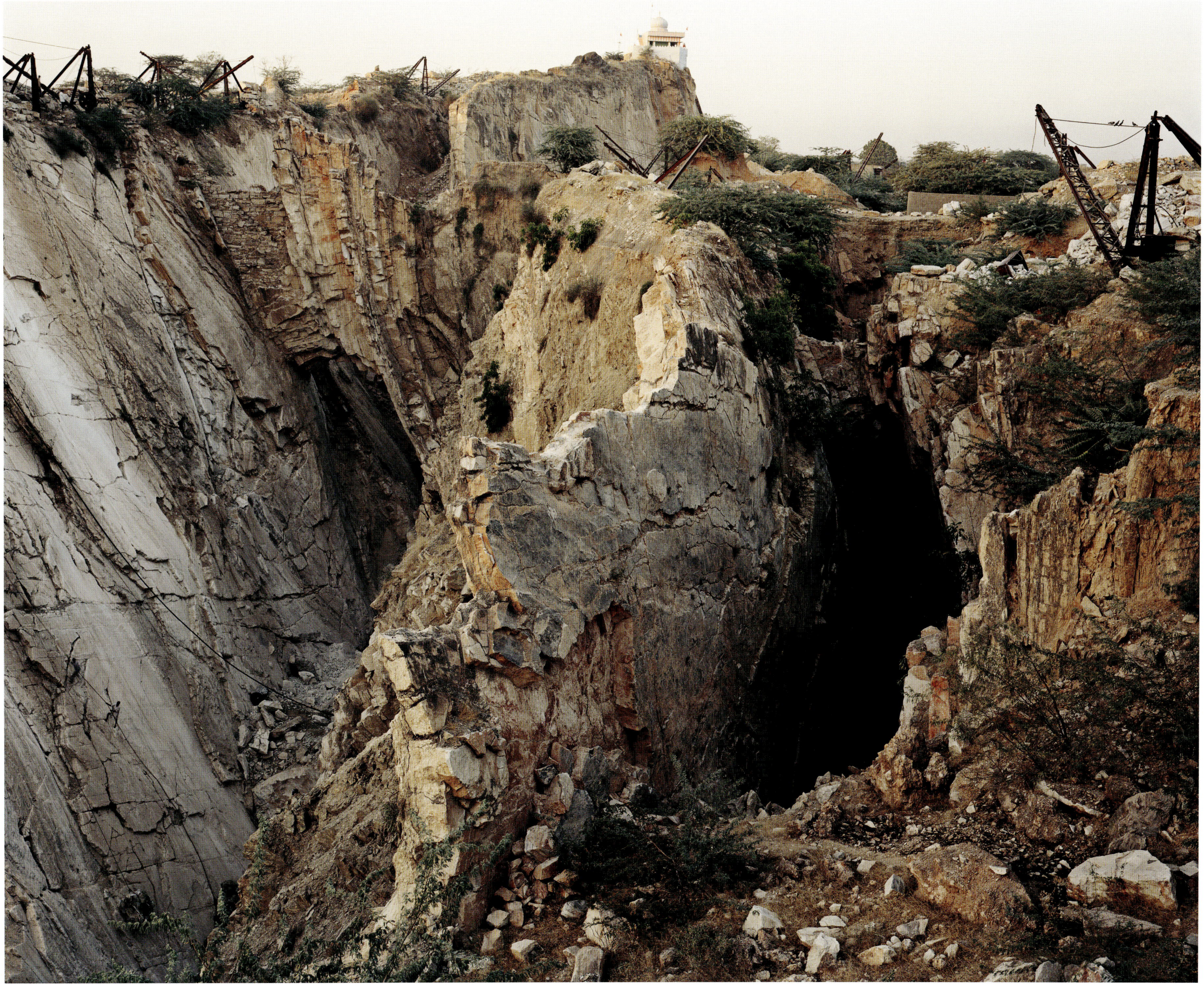

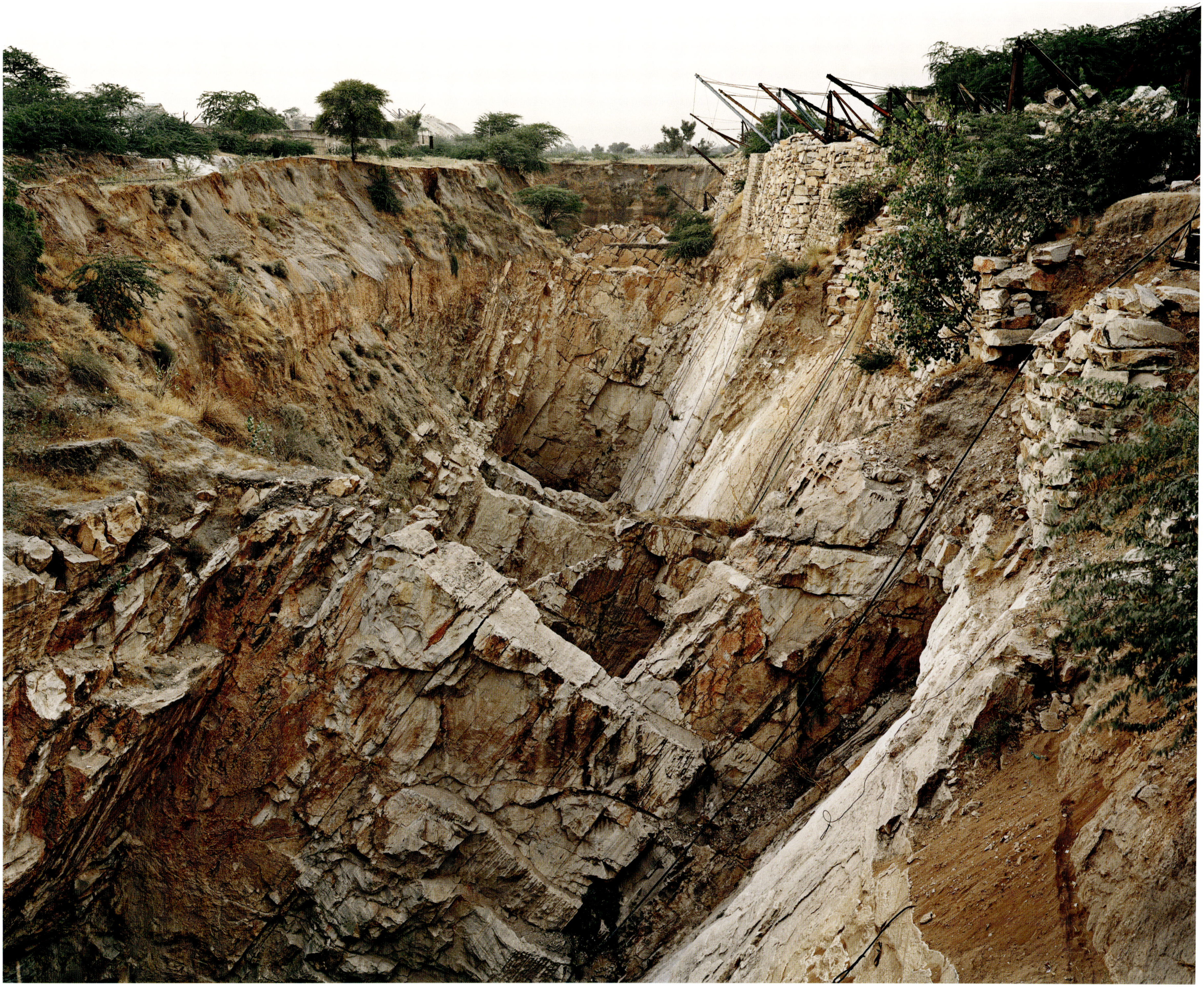

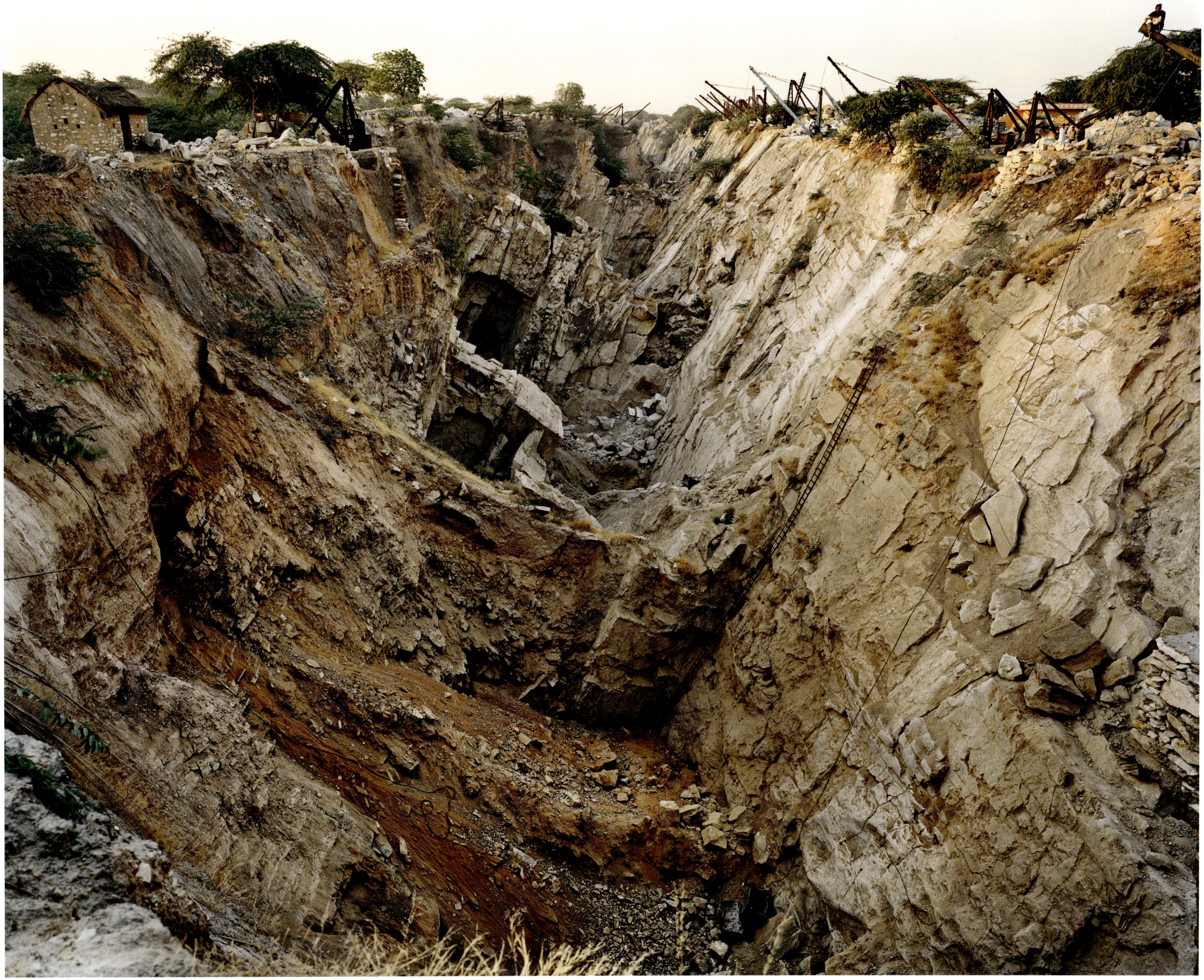

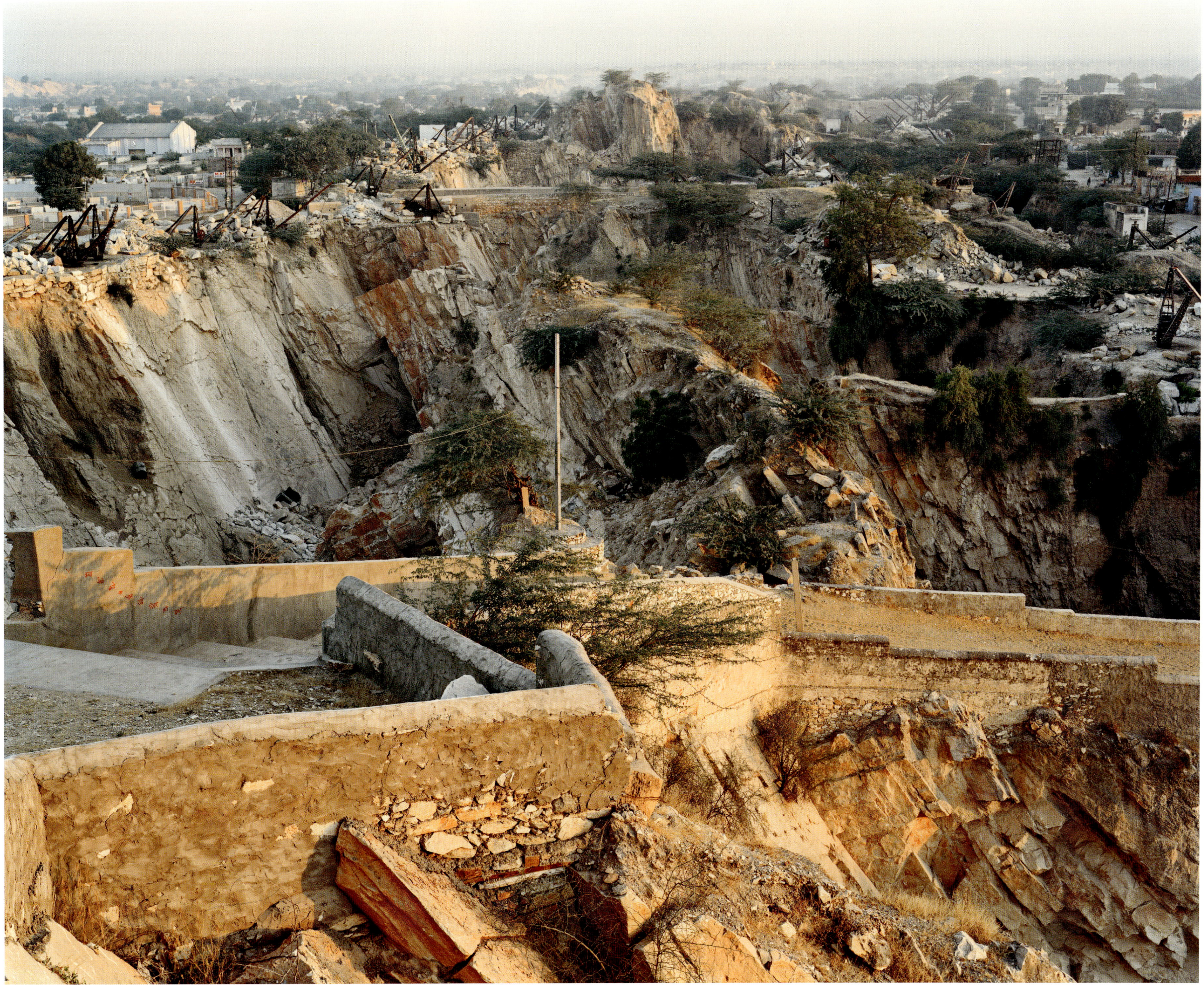

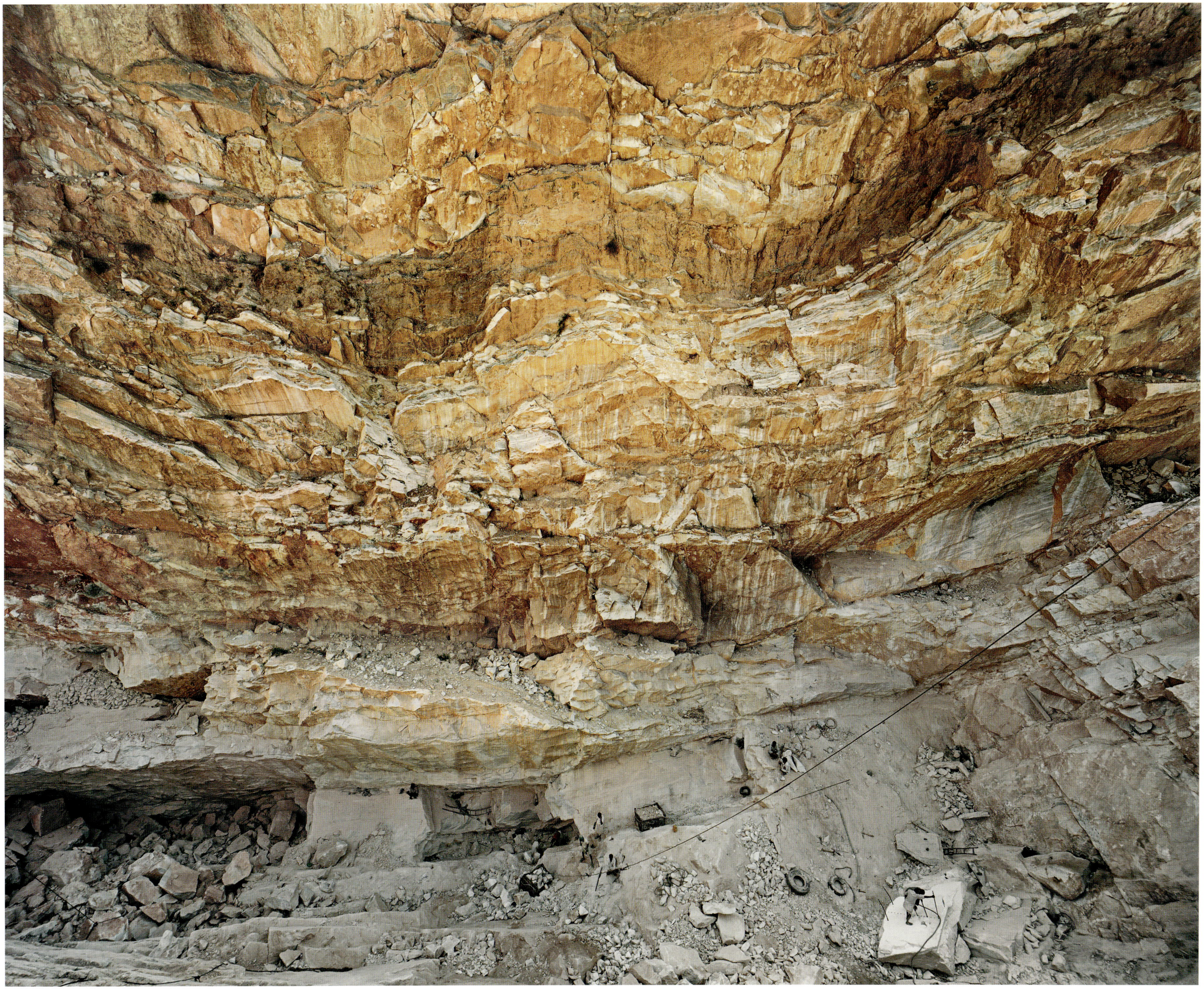

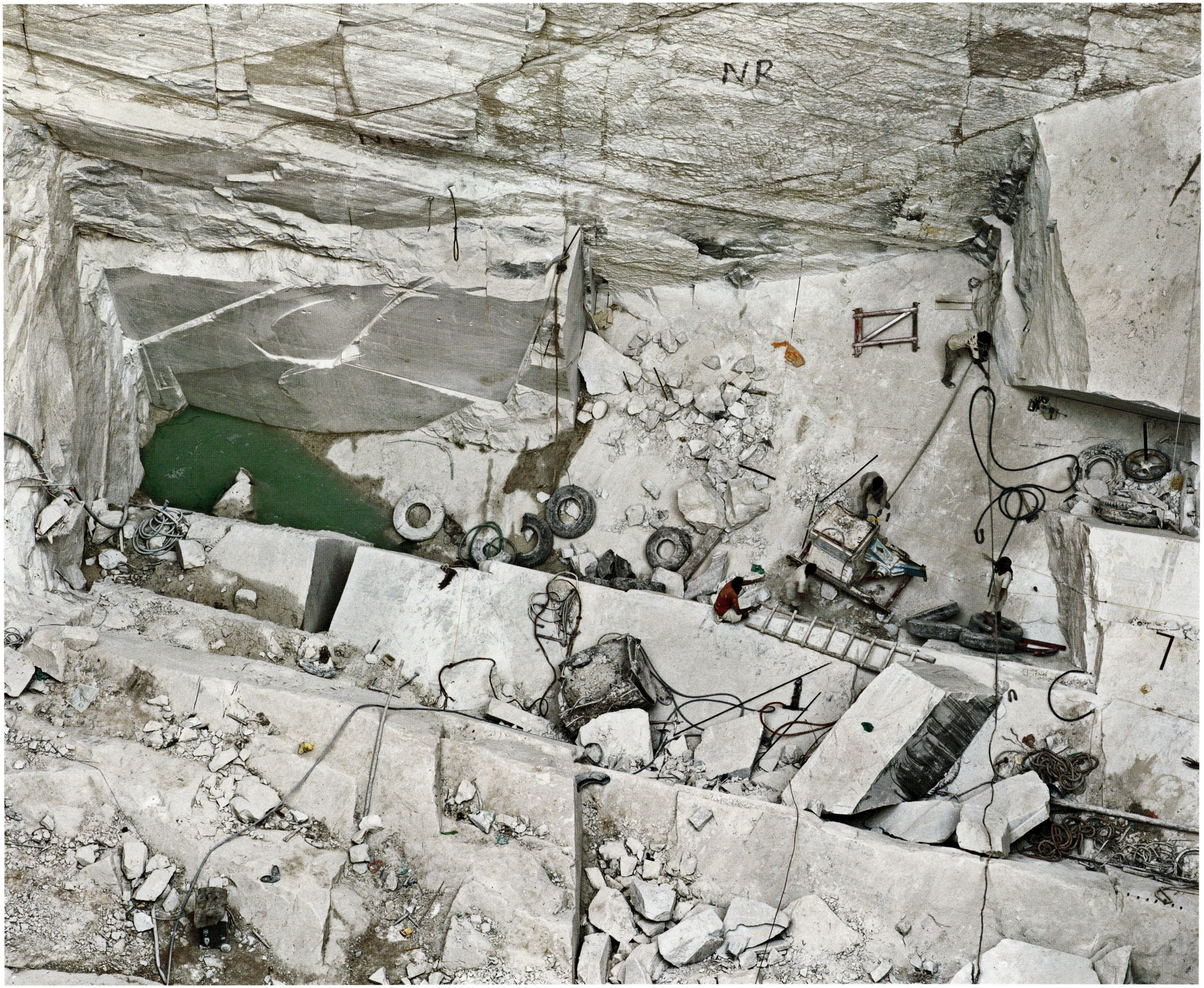
NR

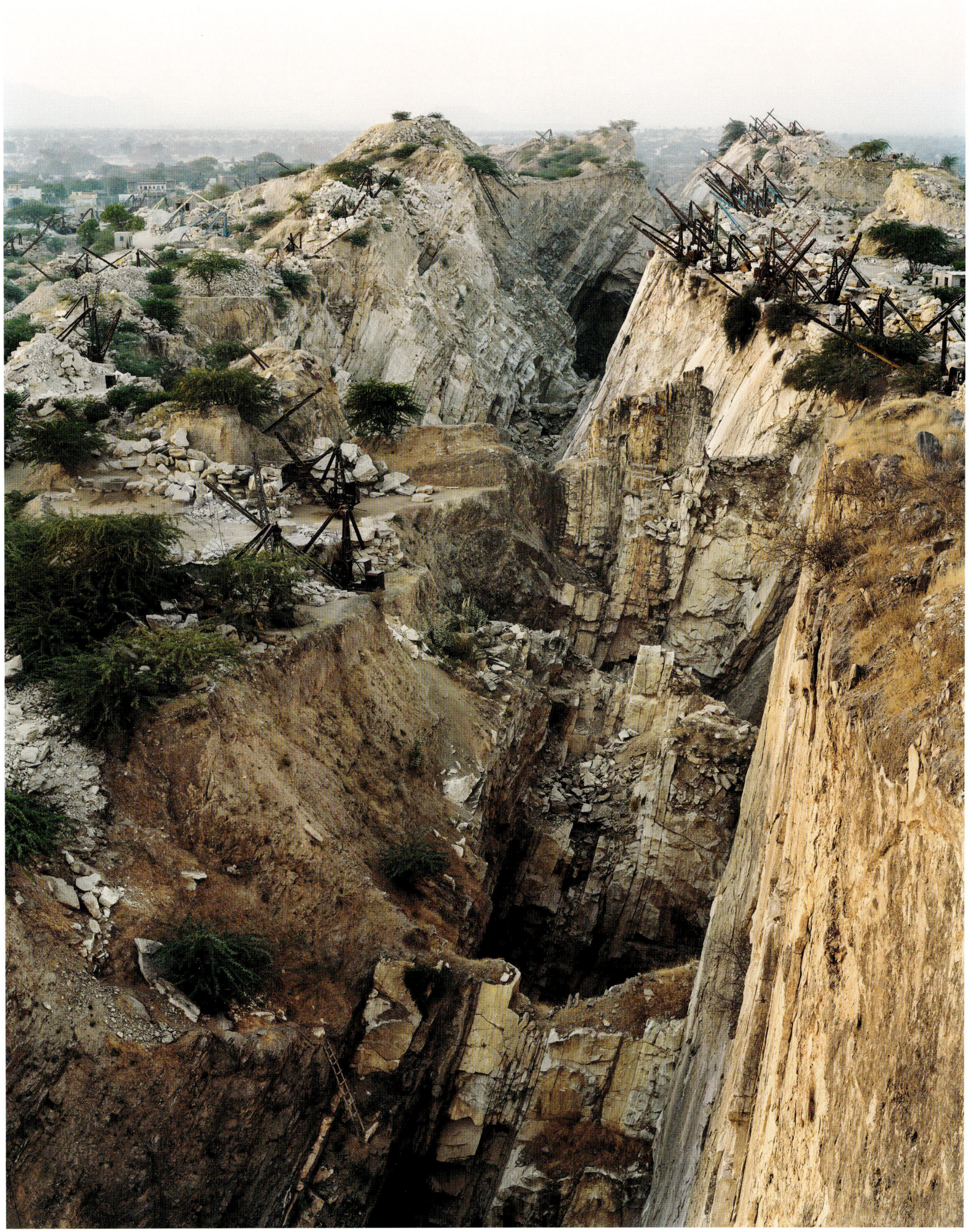

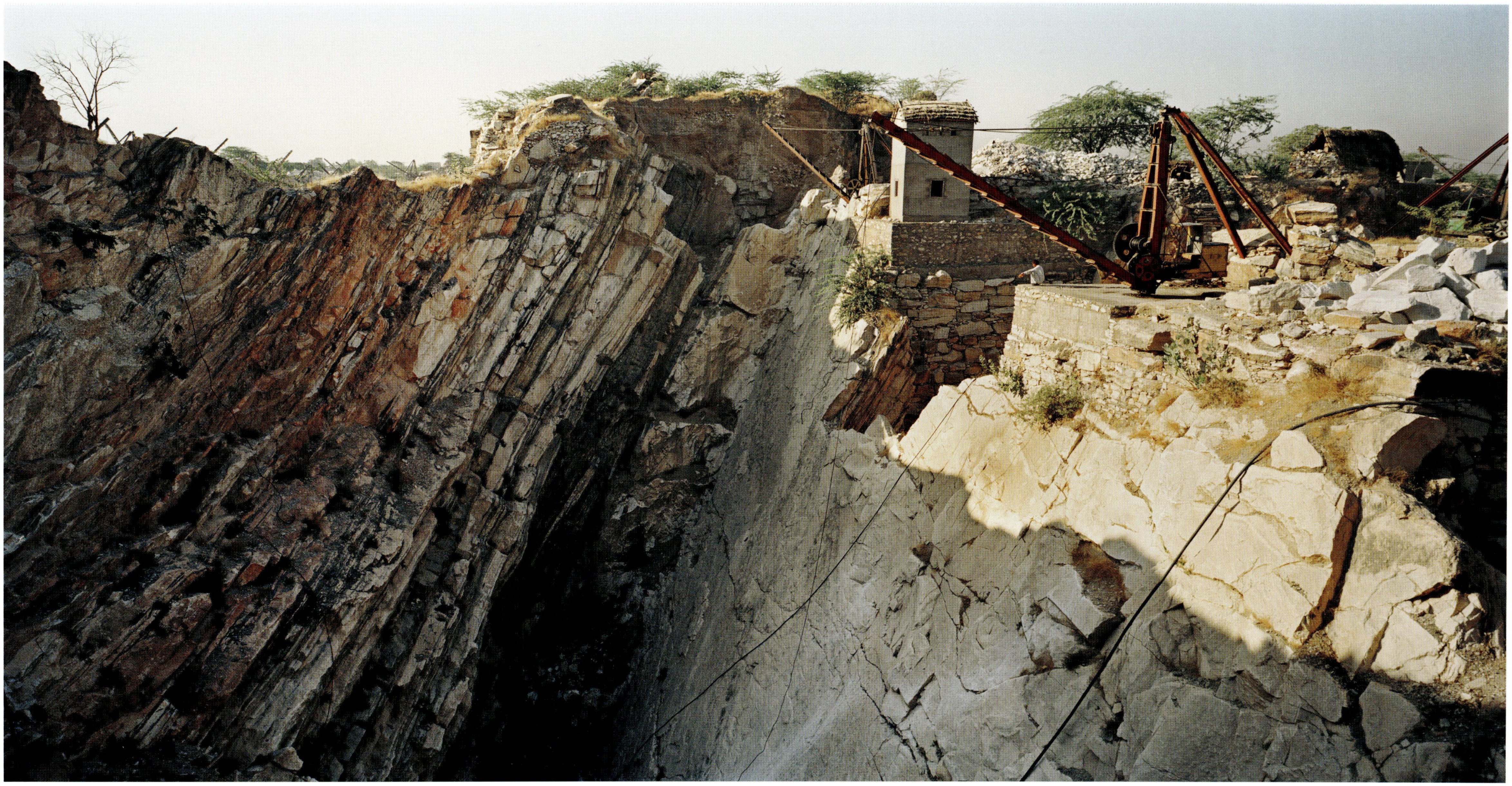

XIAMEN

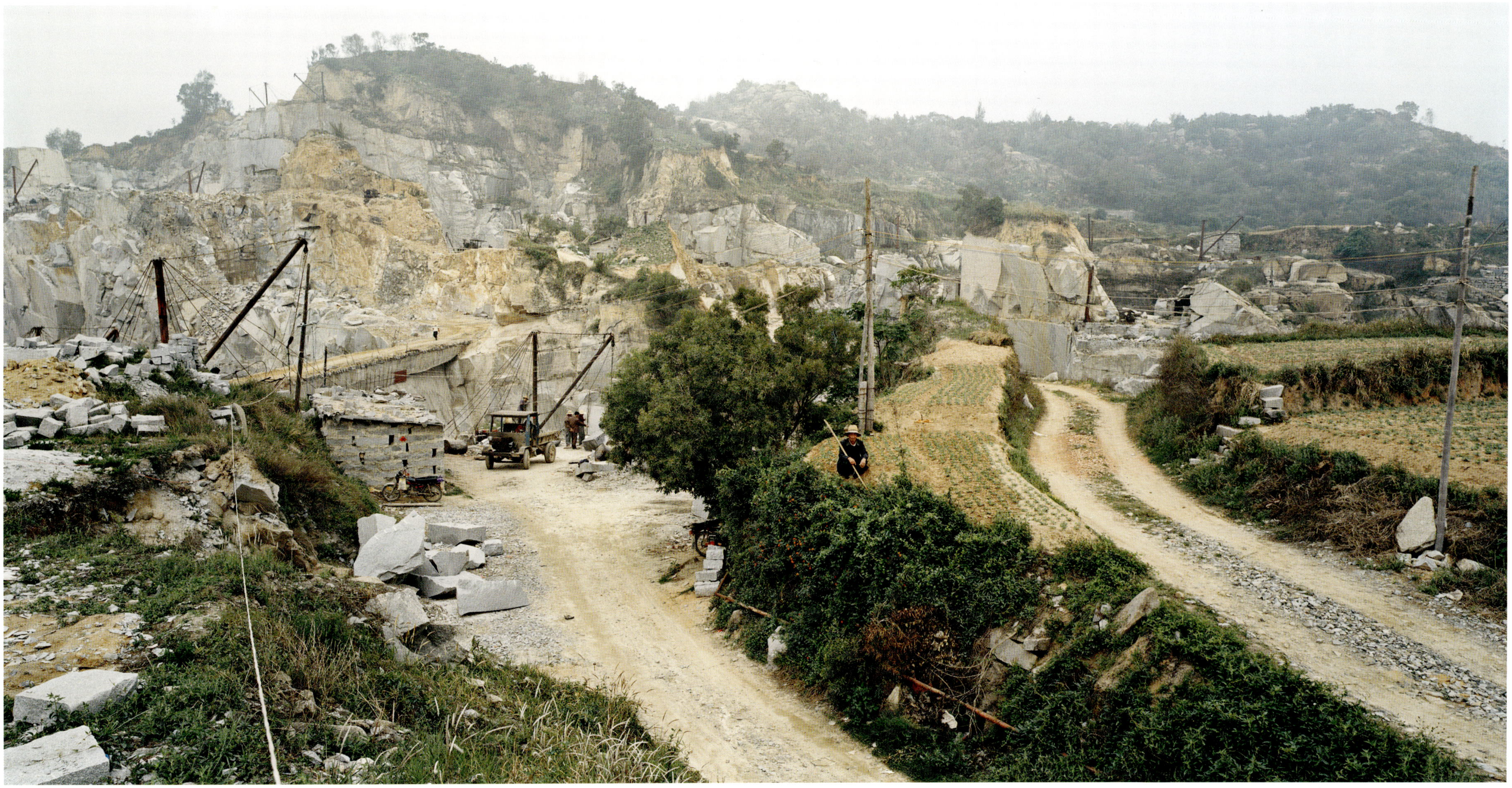

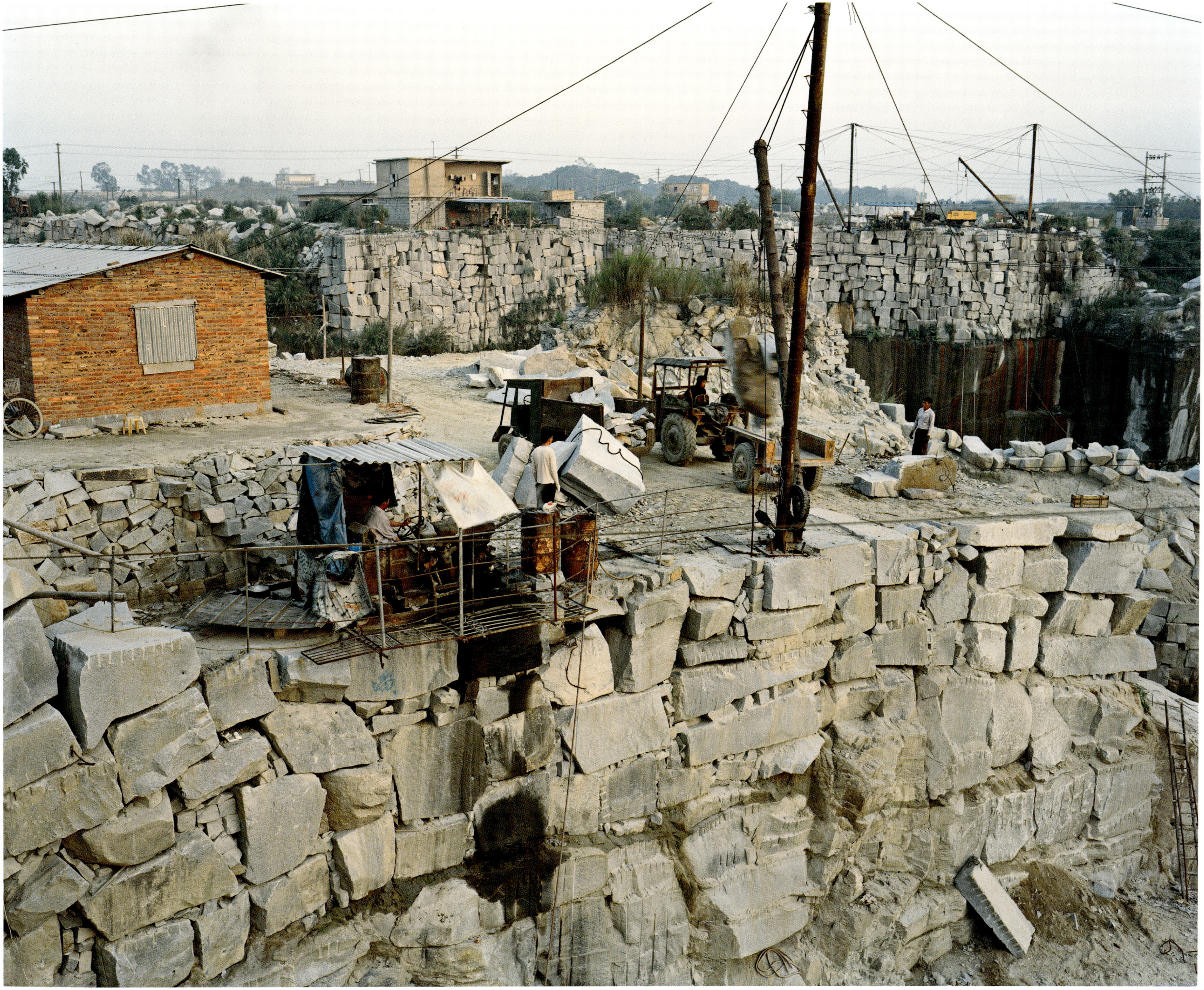

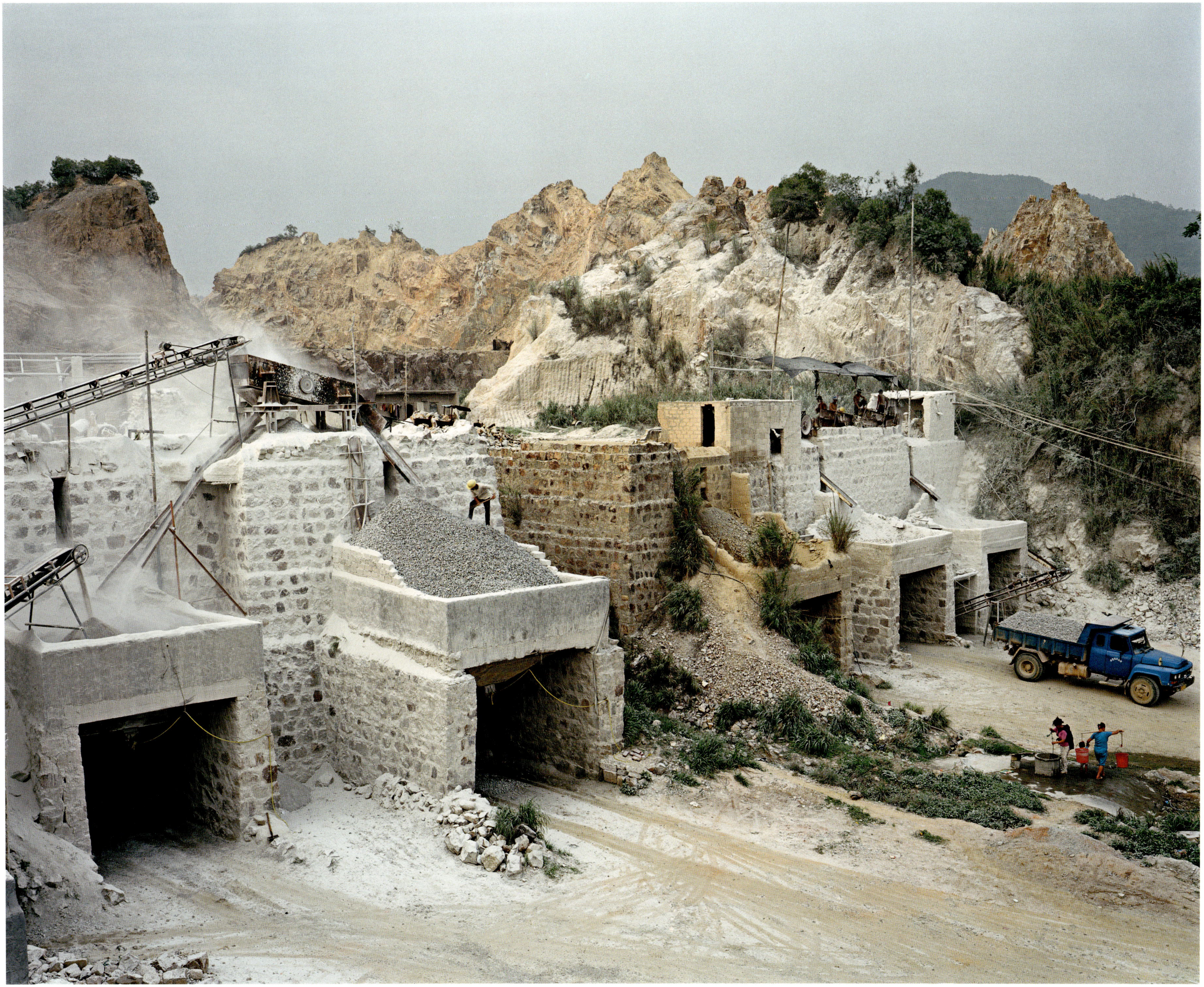

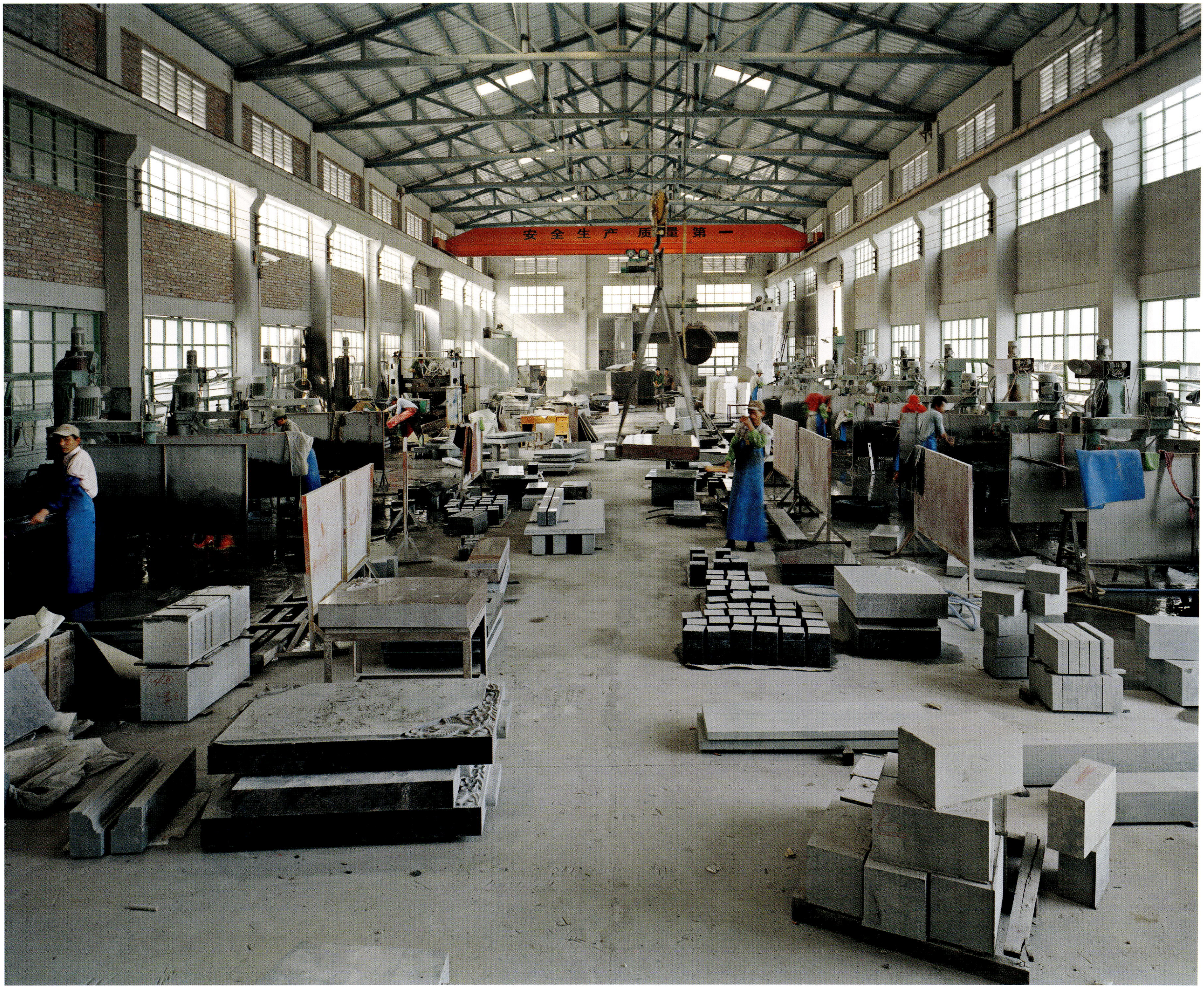
安全生产 质量第一

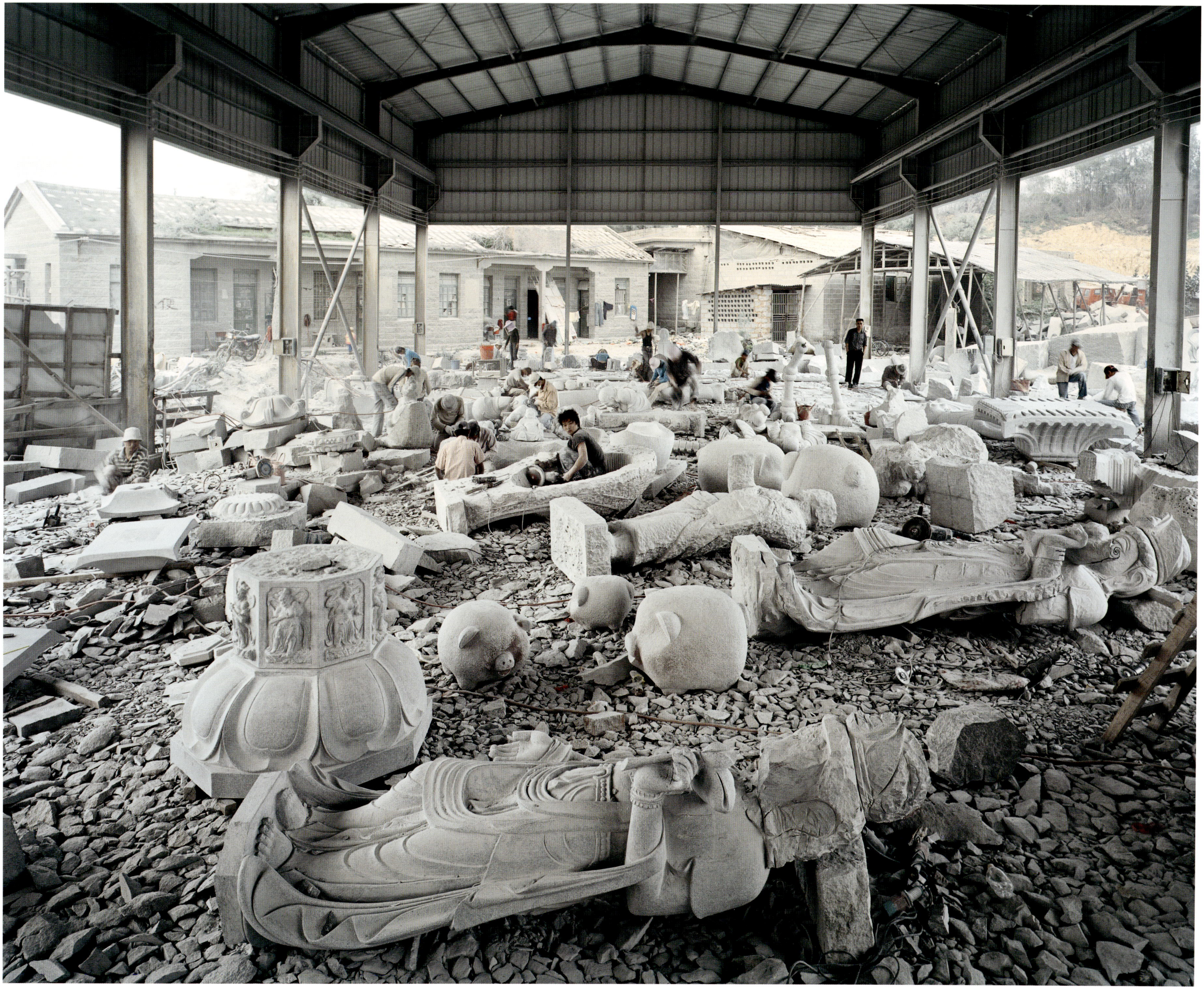

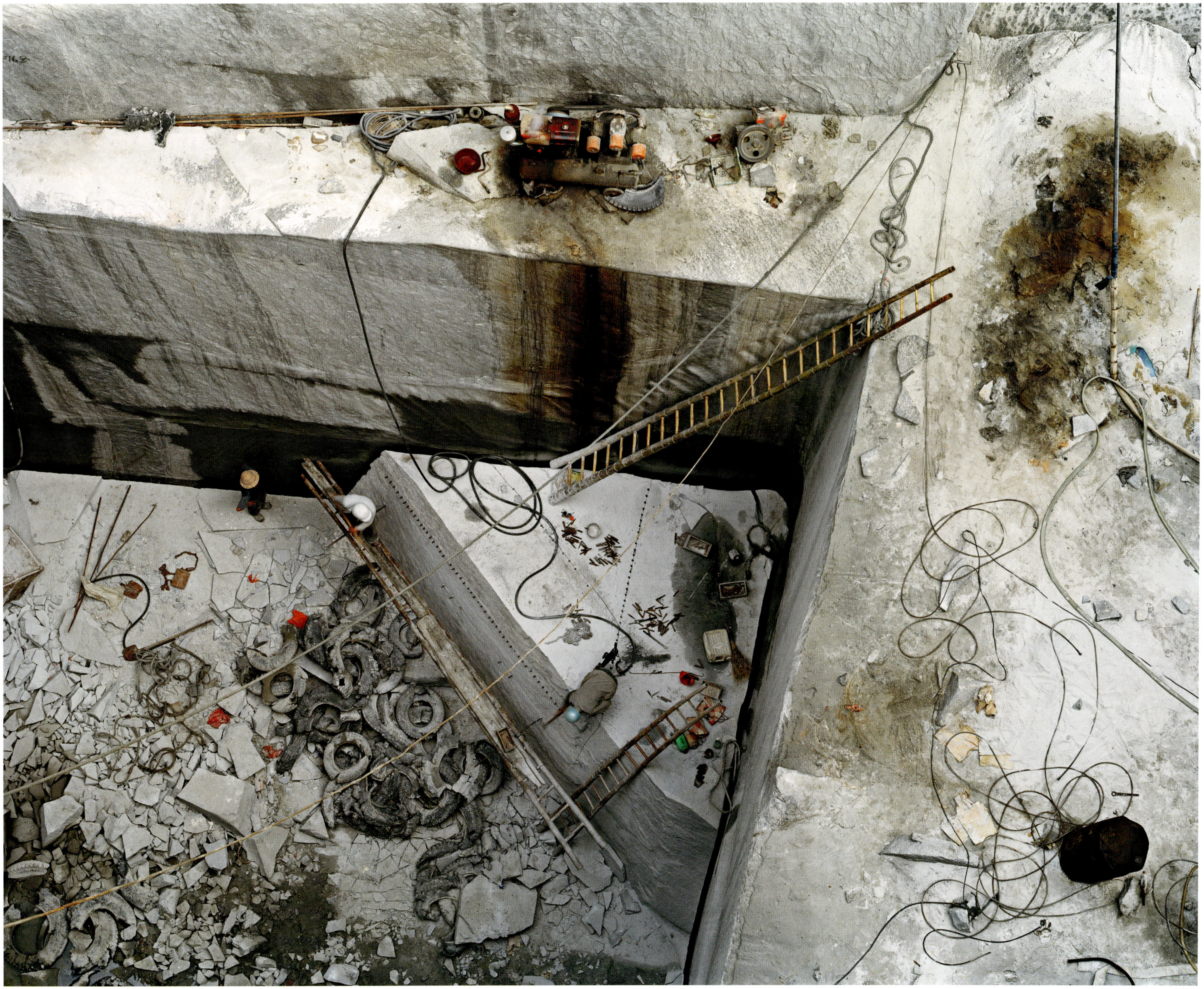

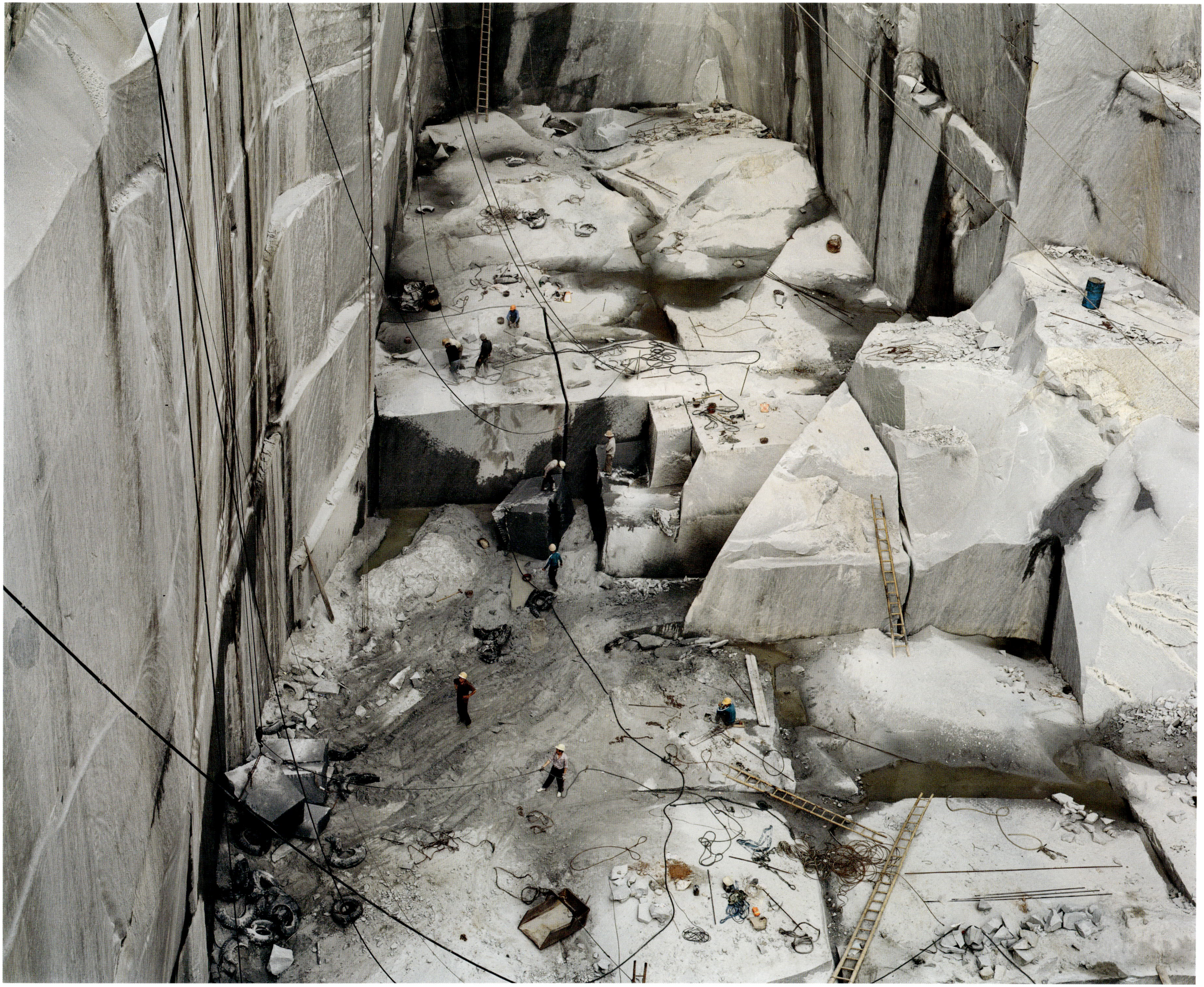

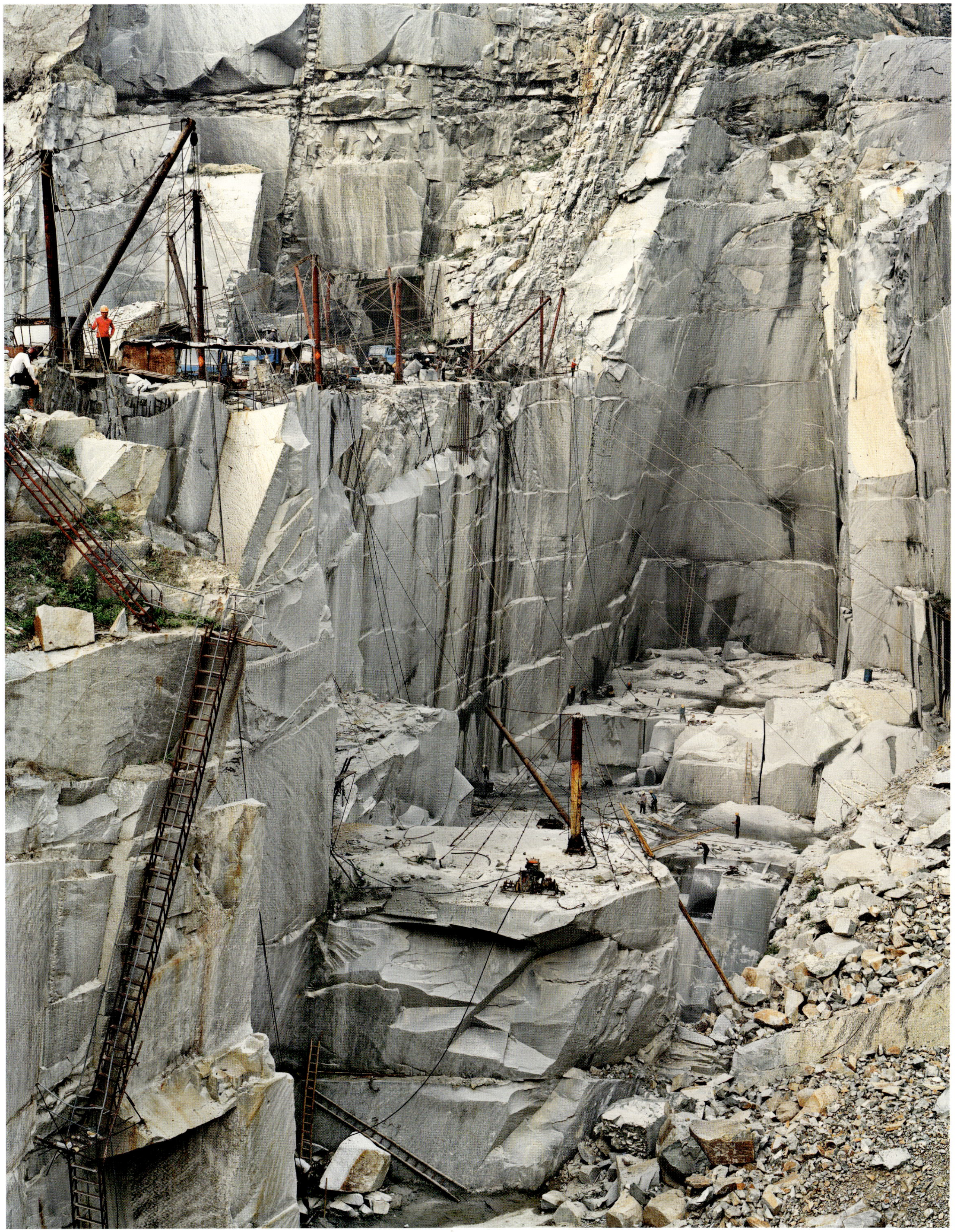

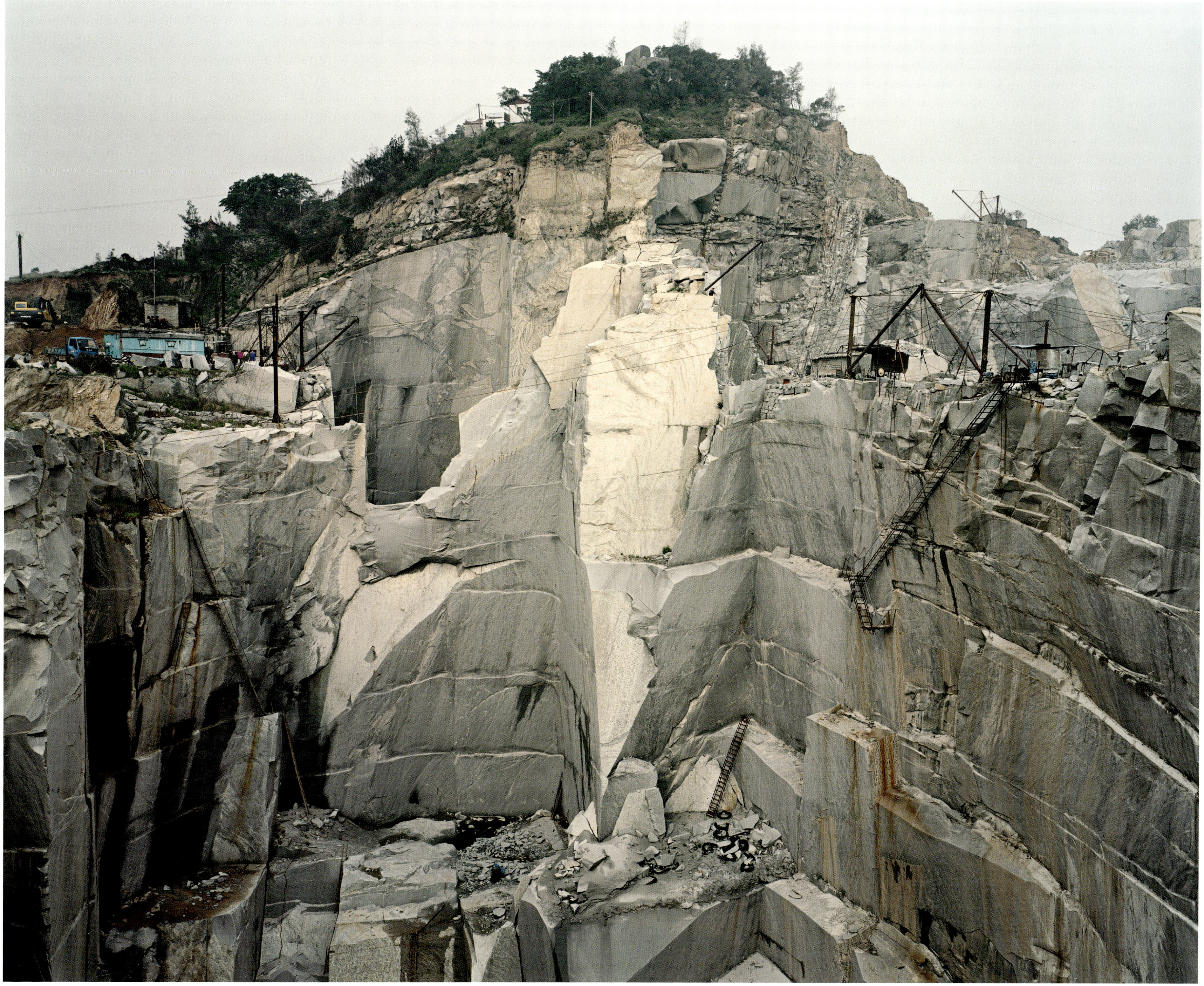

IBERIA

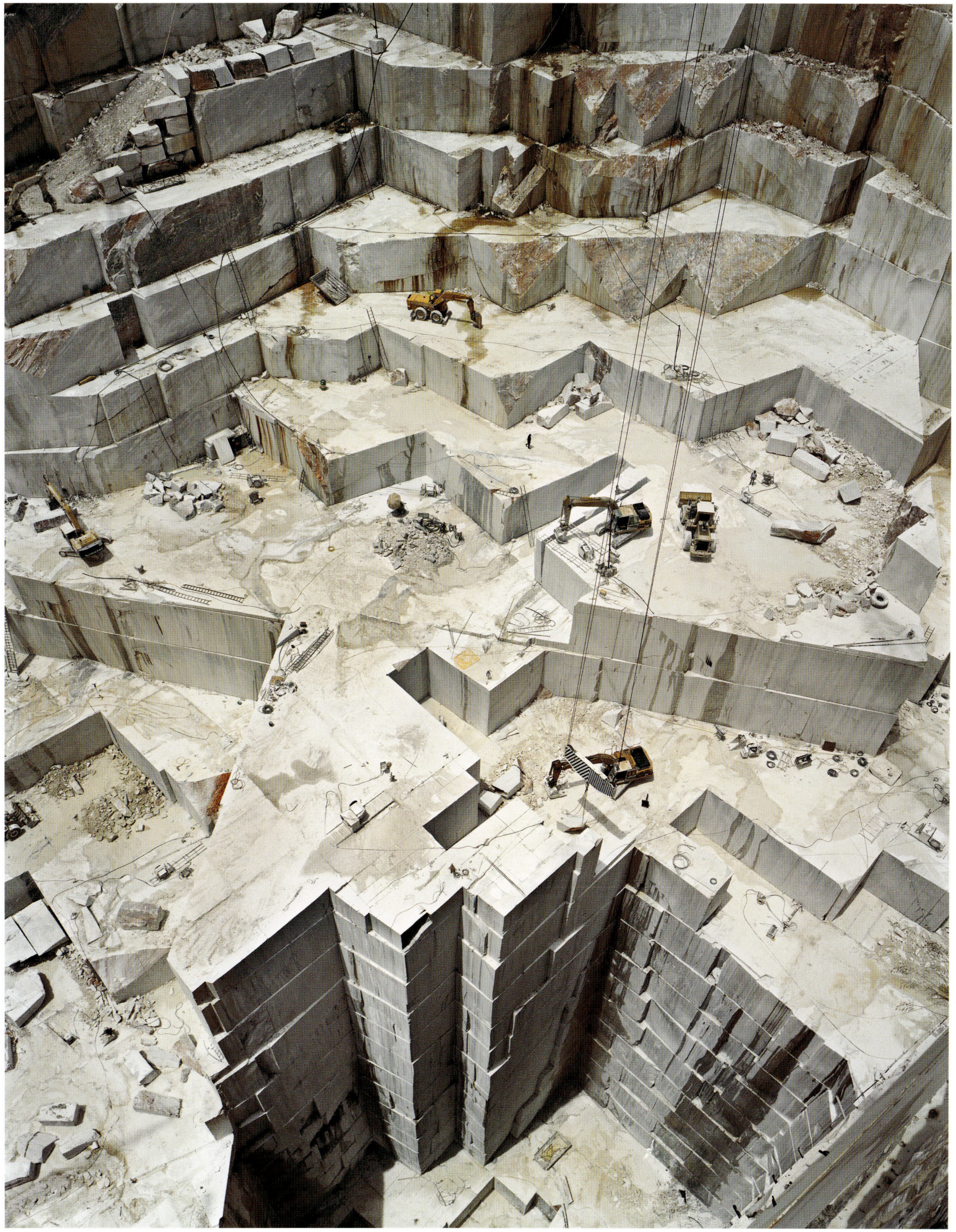

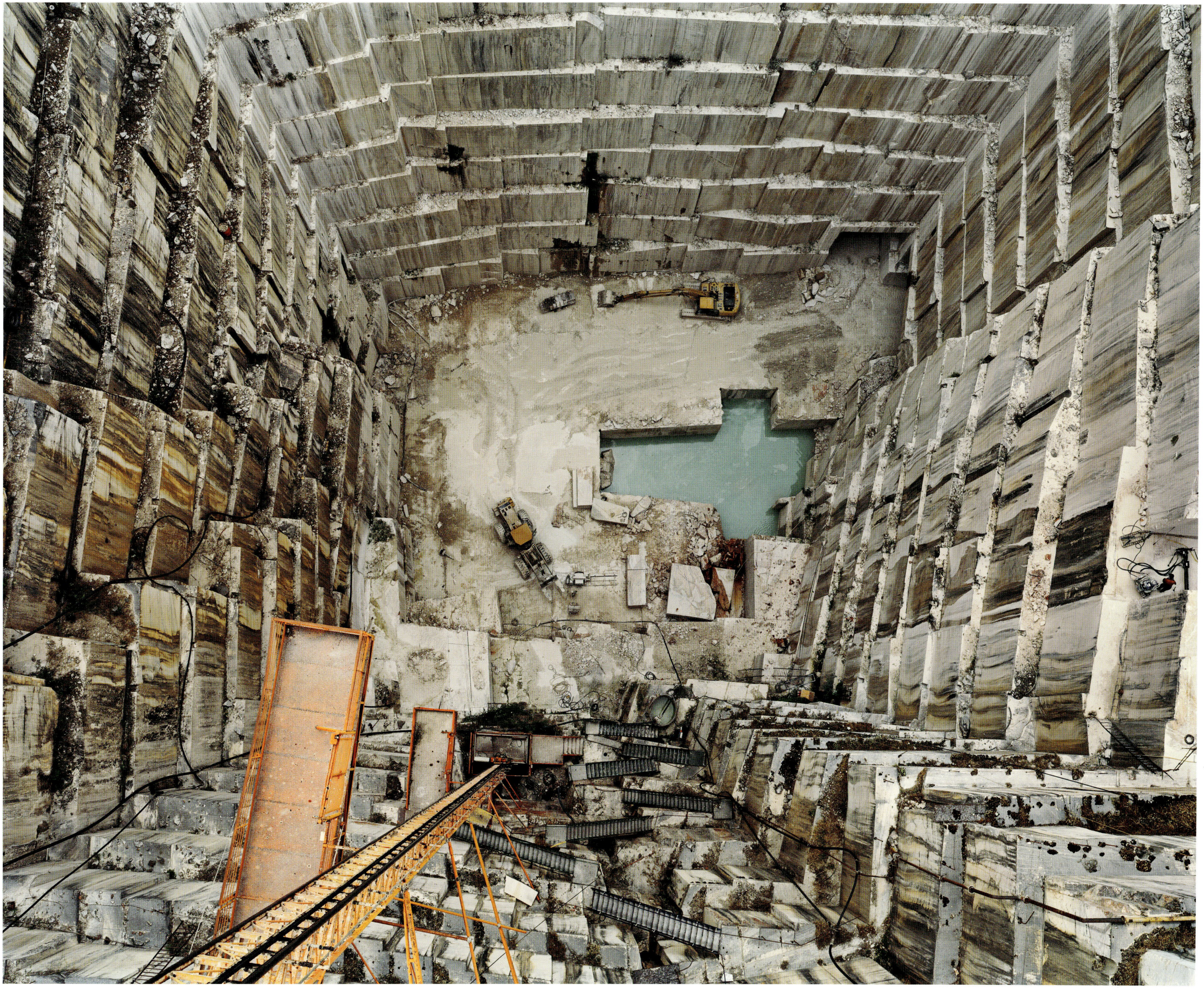

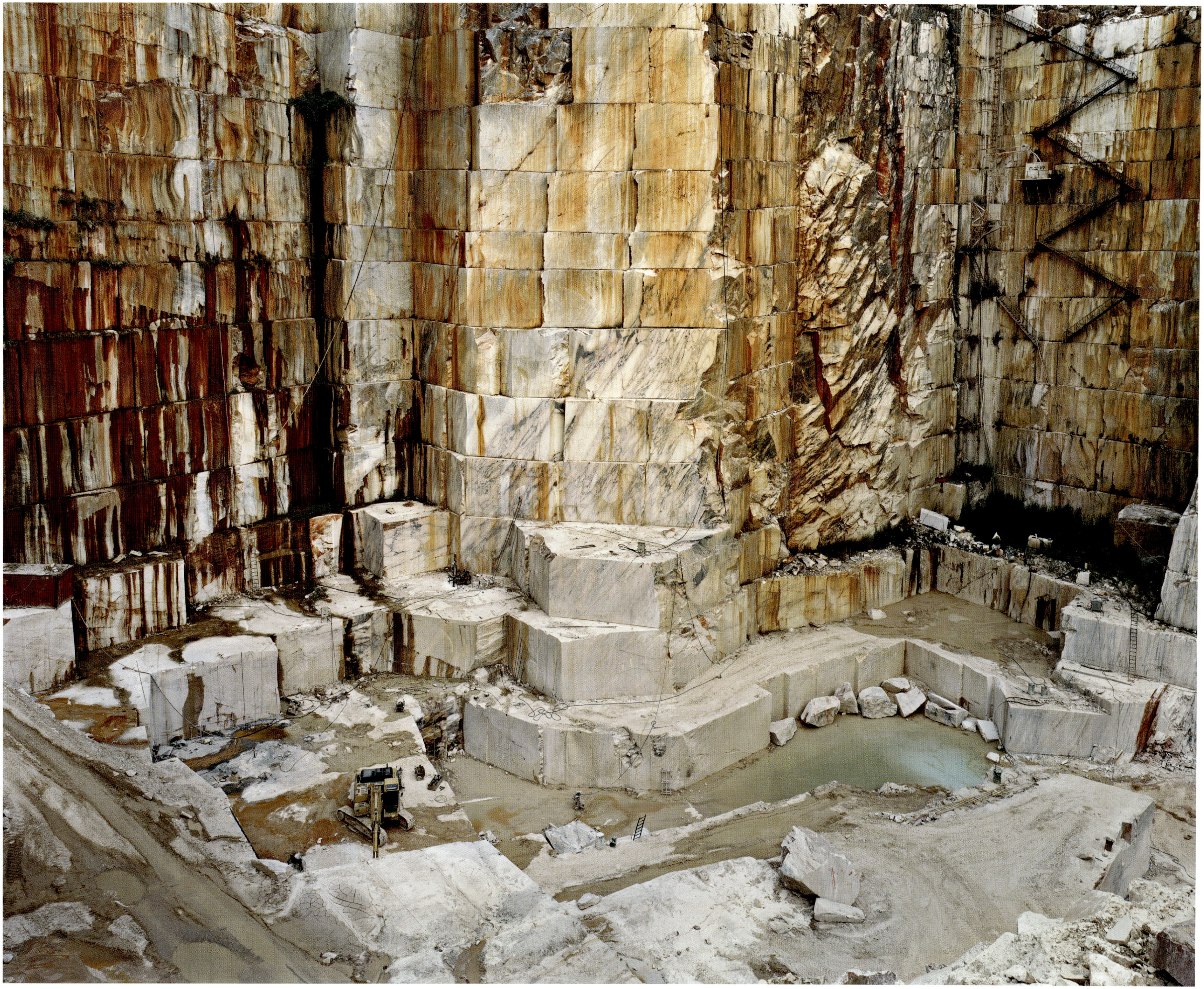

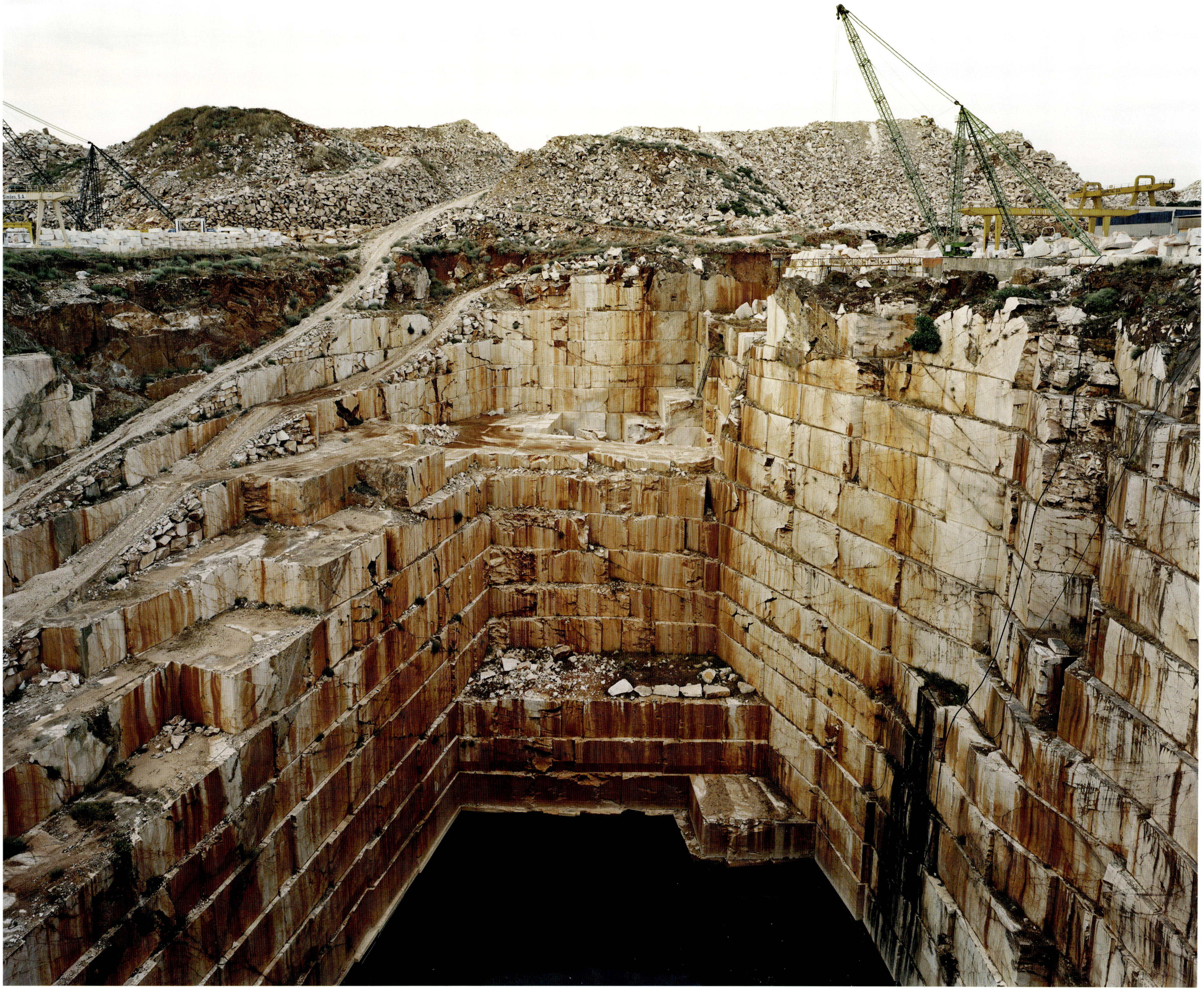

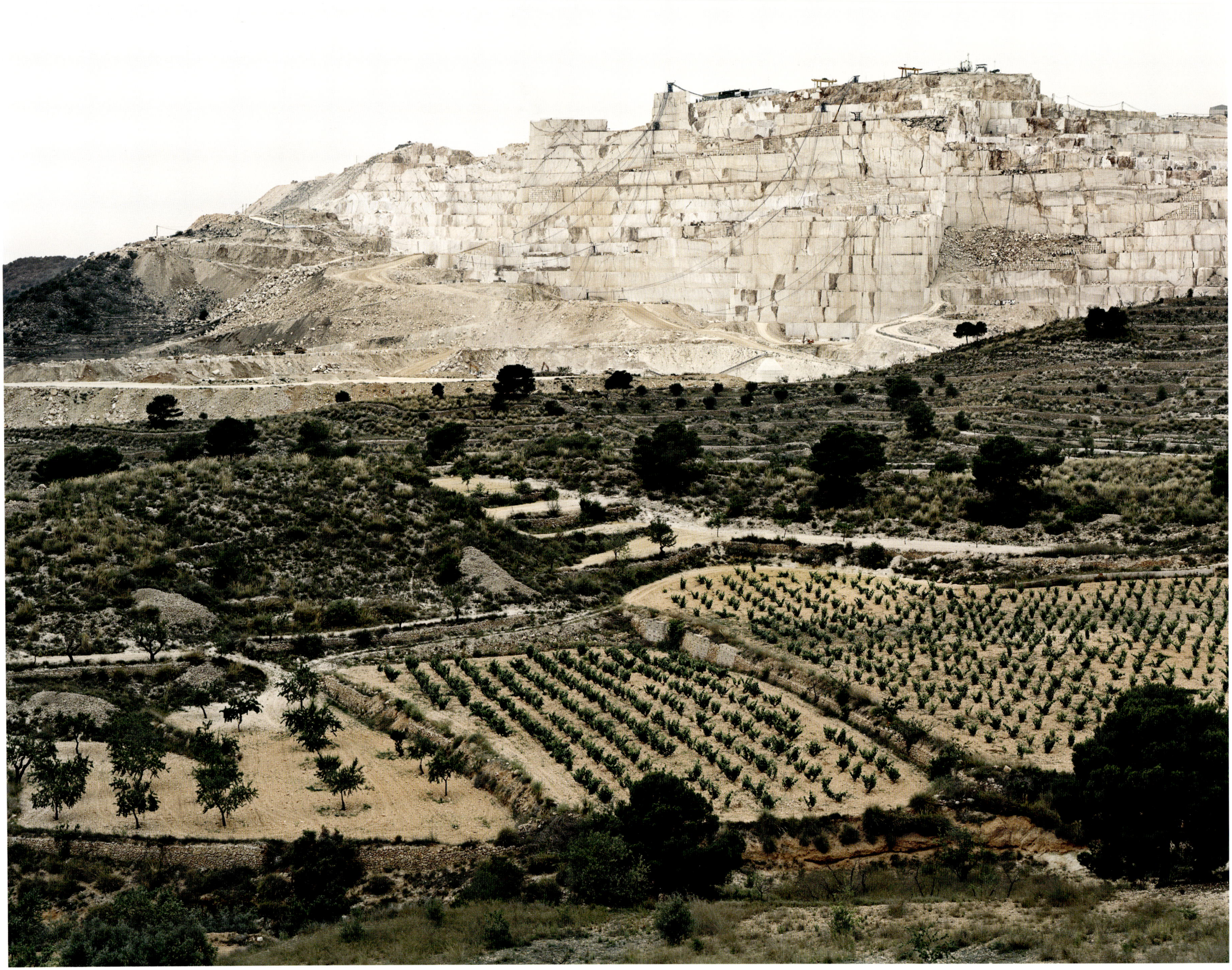

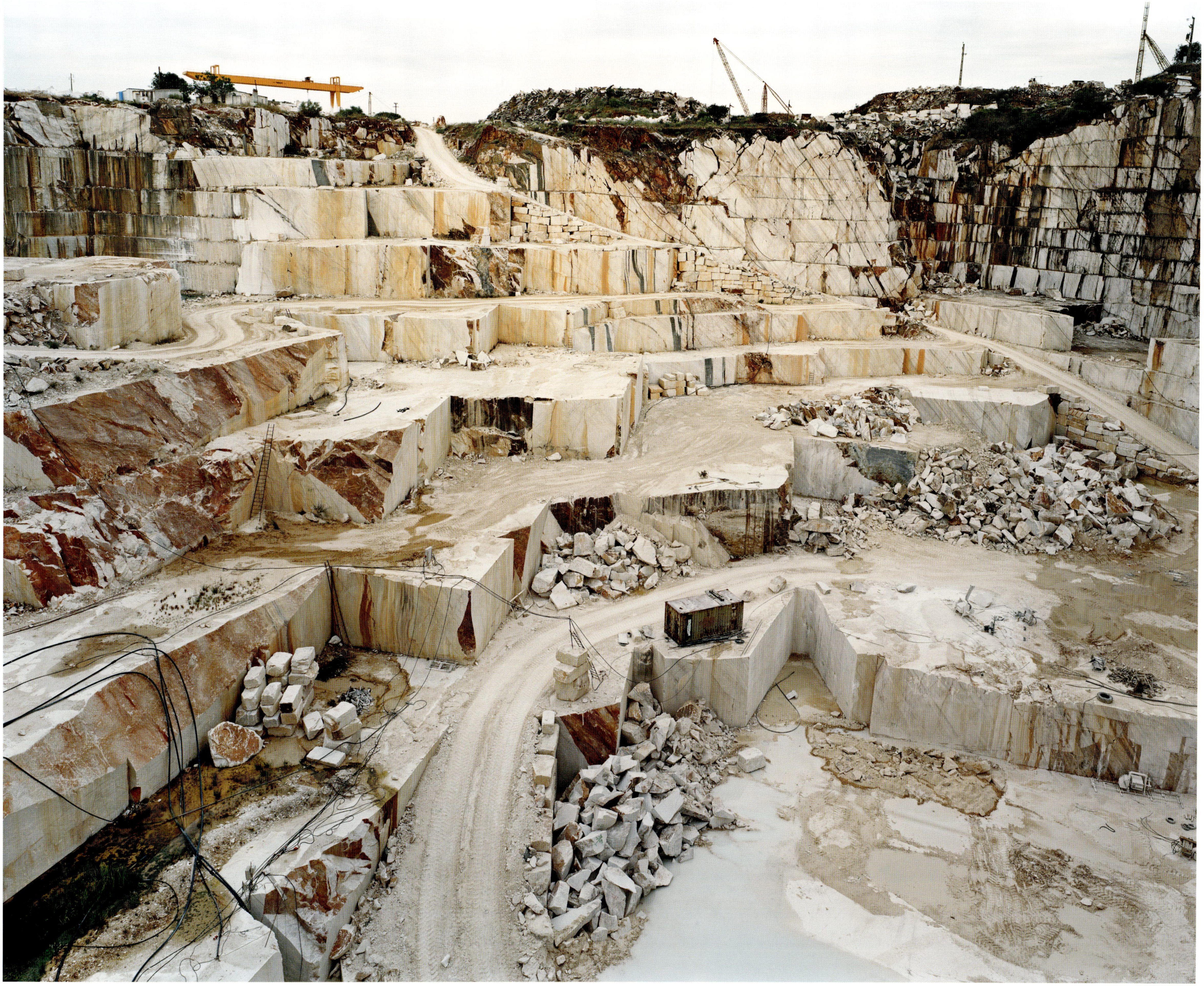

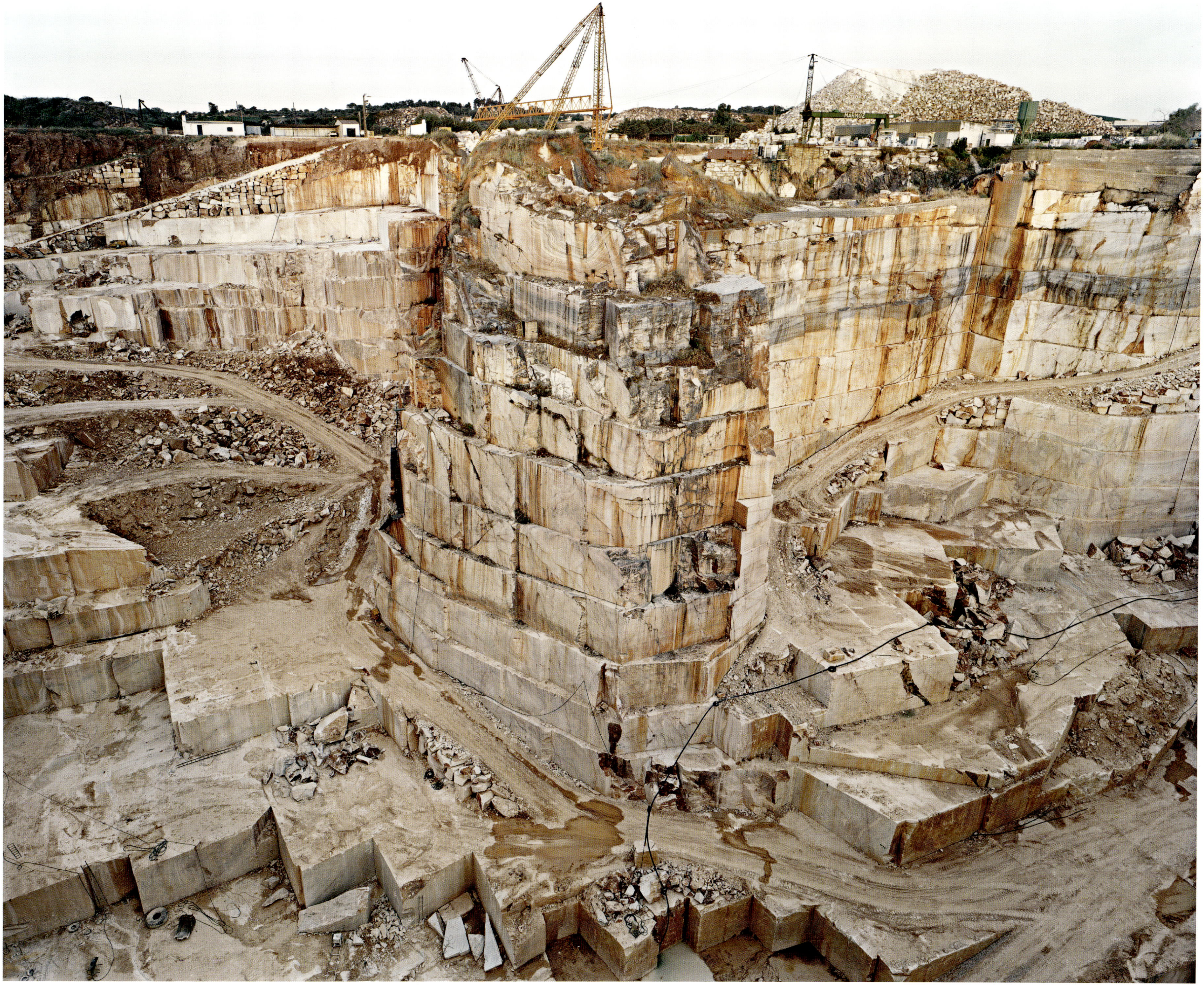

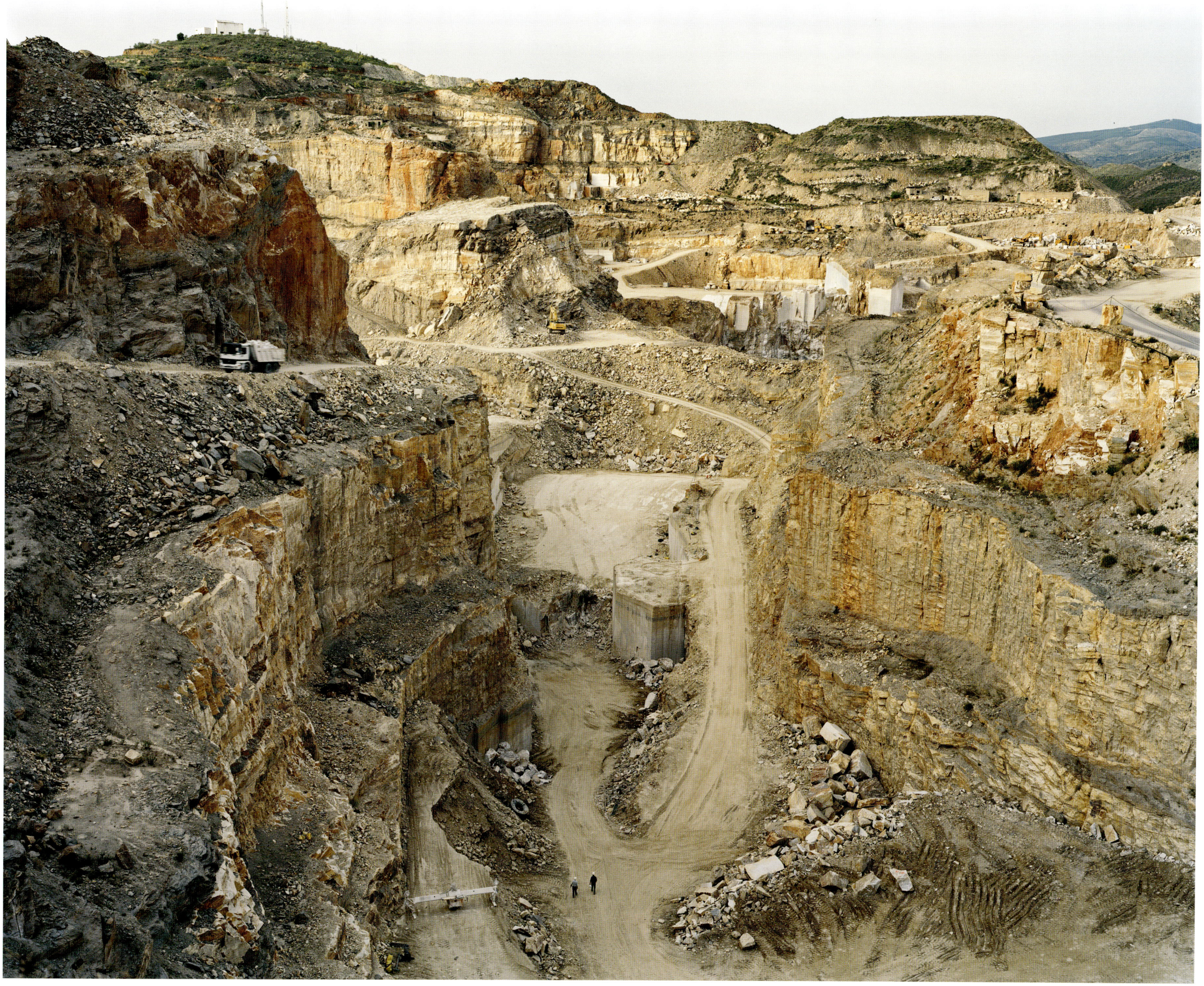

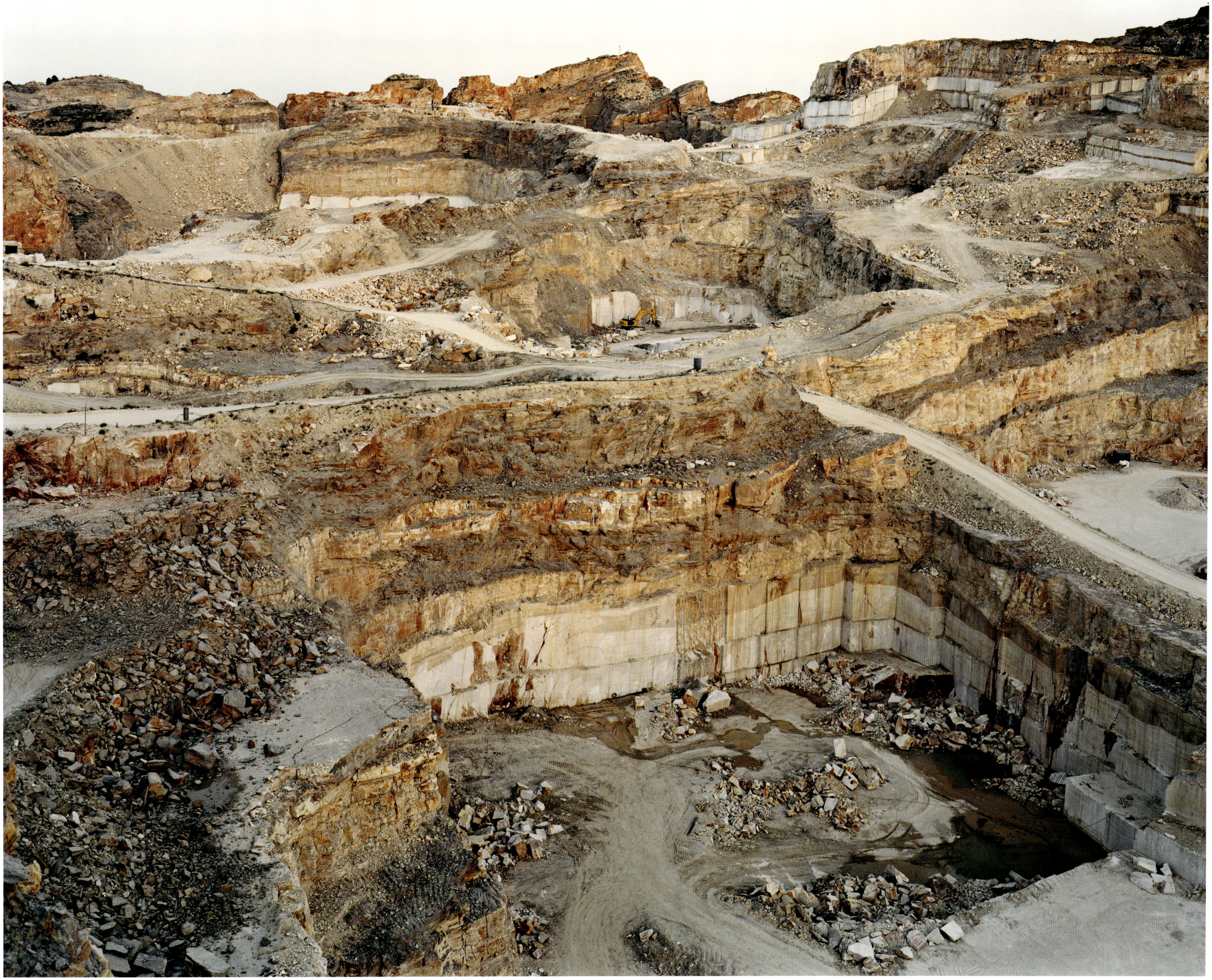

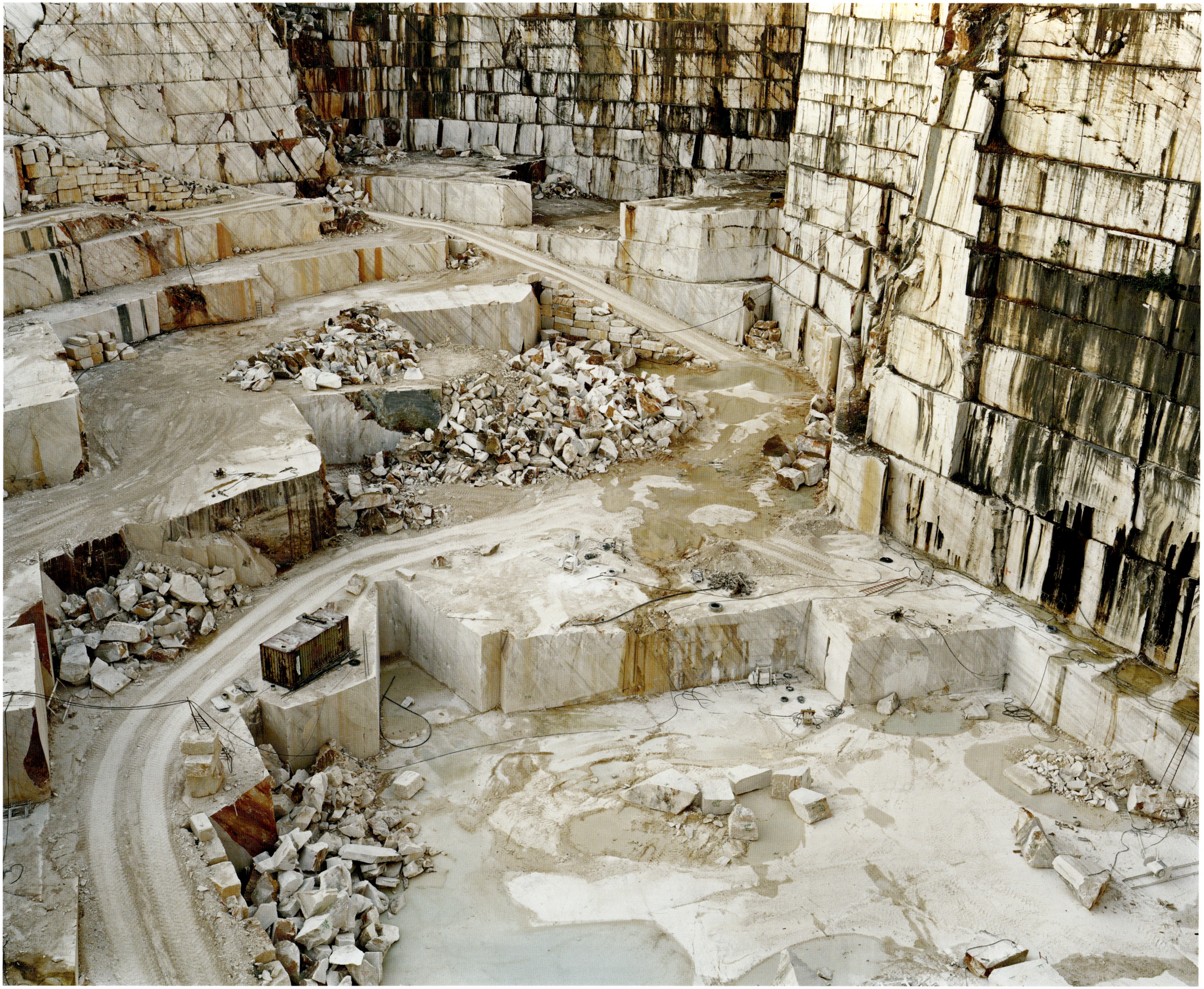

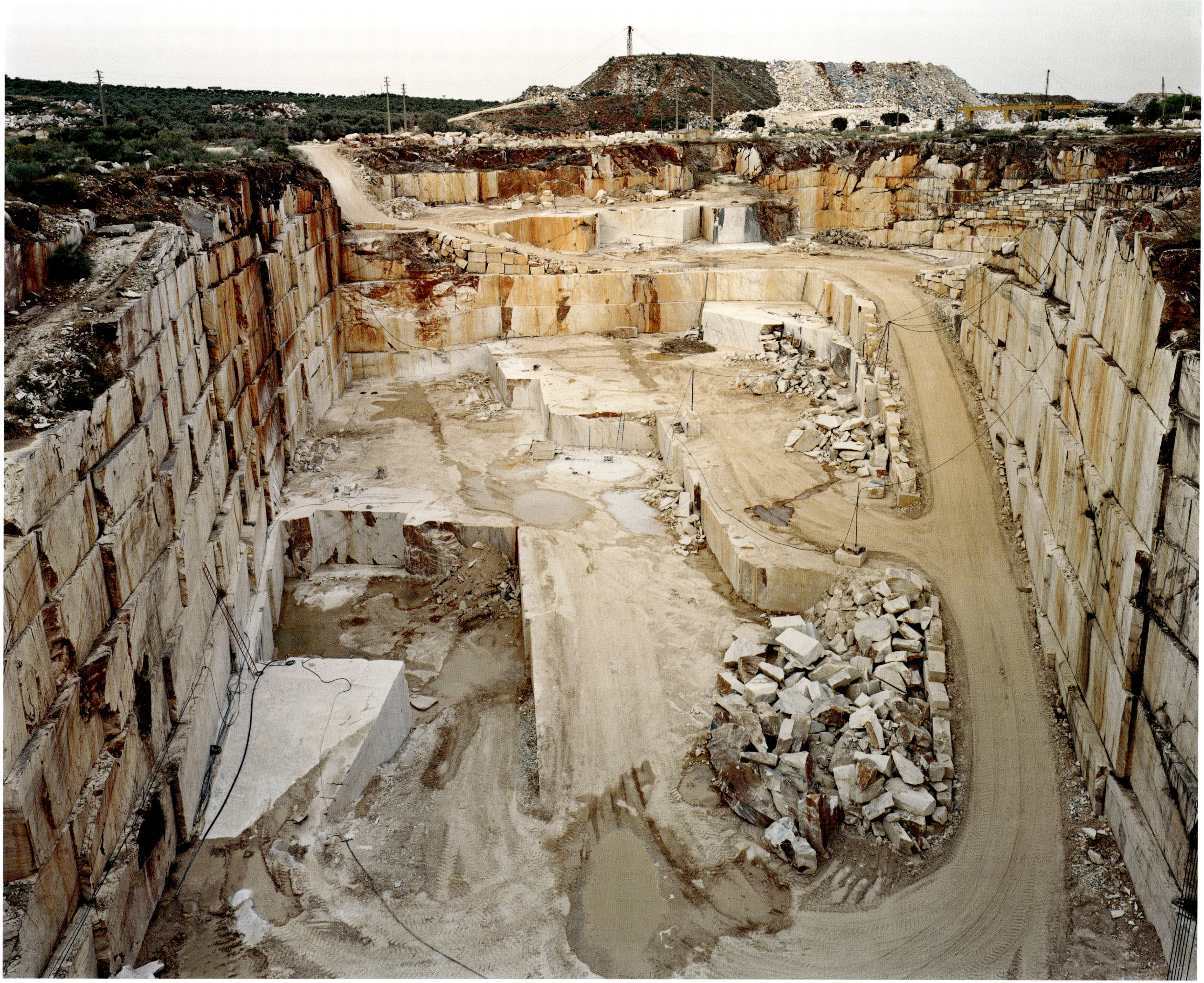

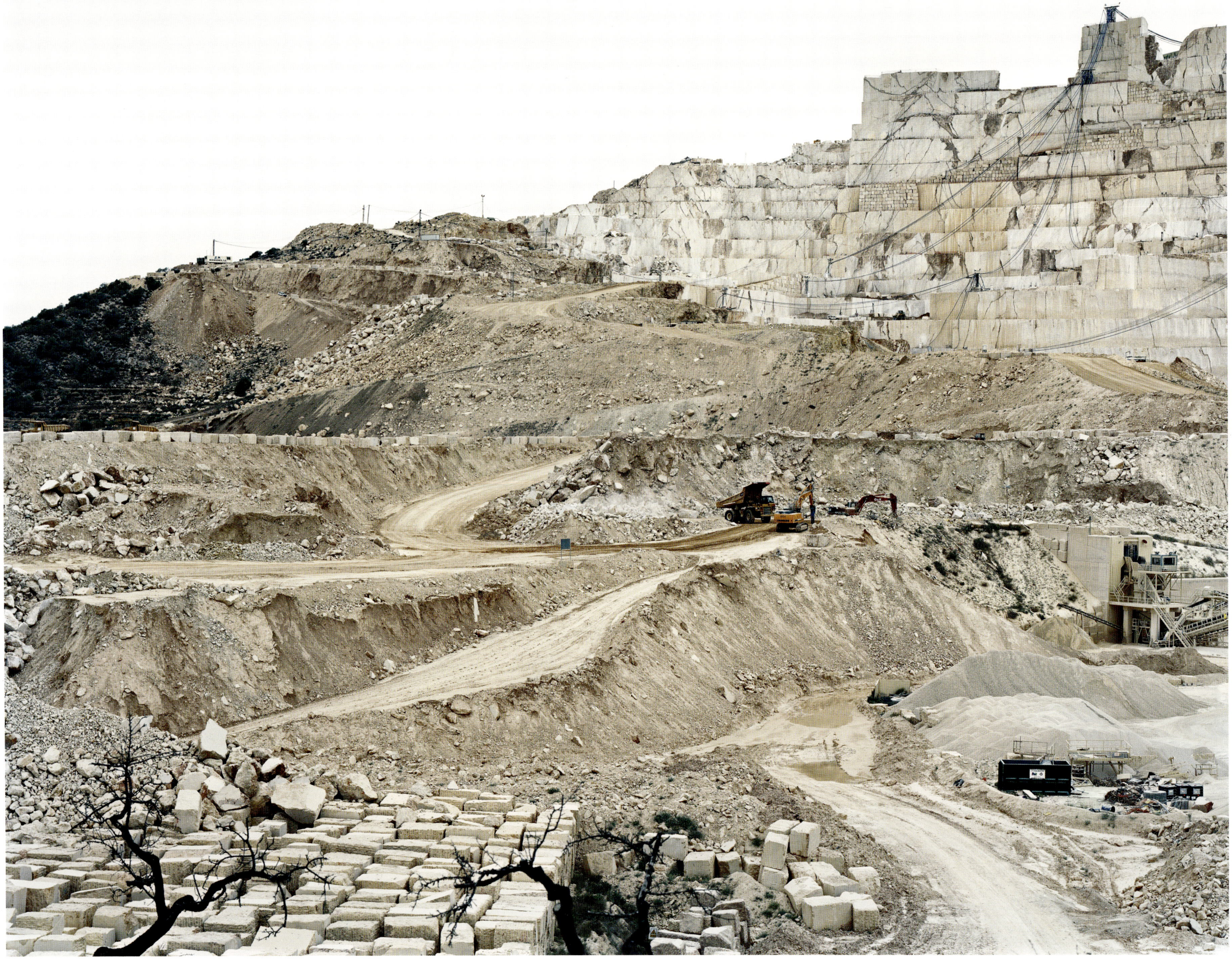

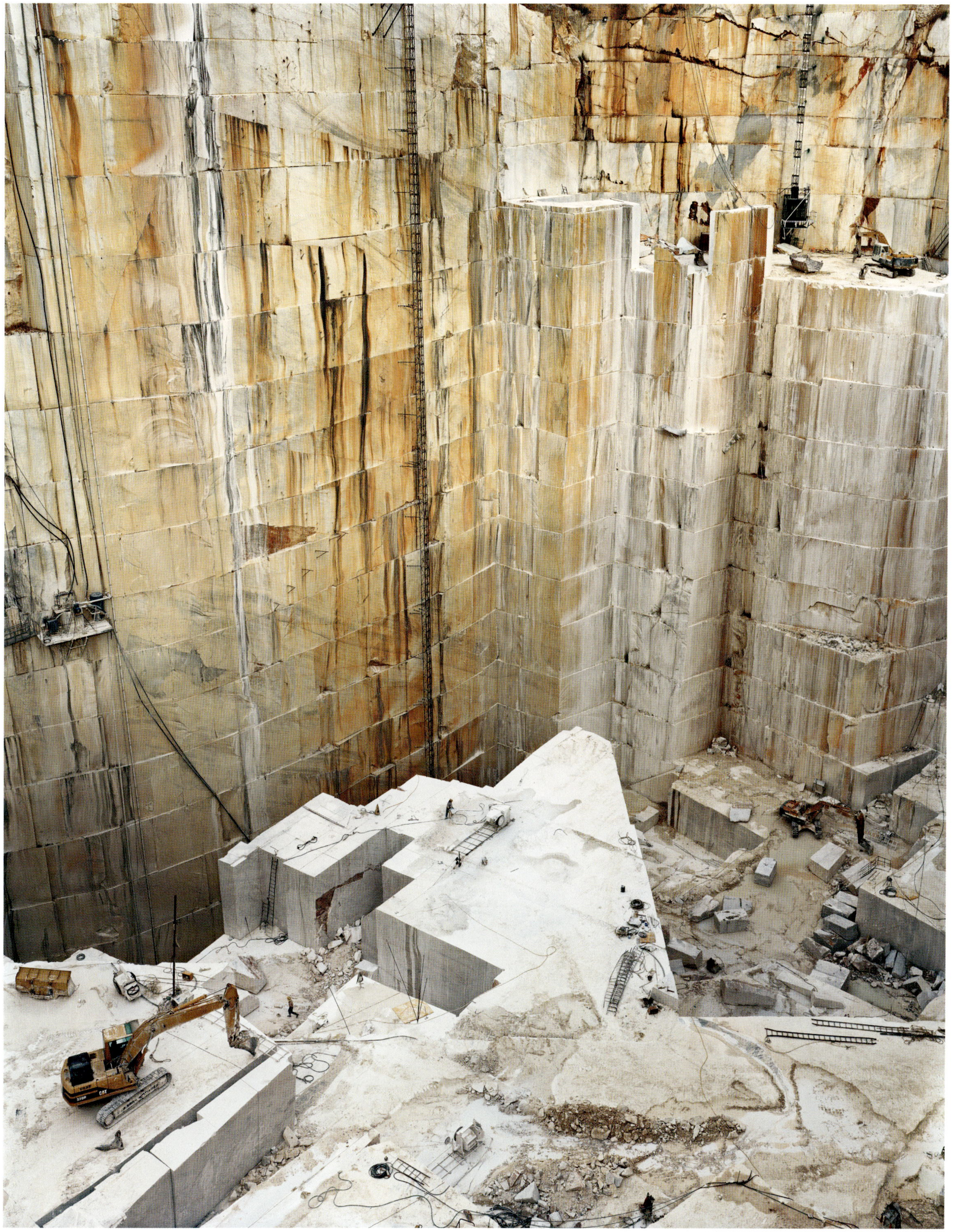

CAPTIONS

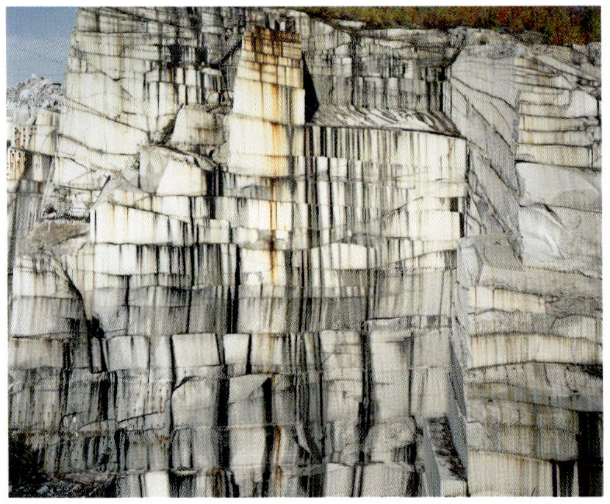

29

Rock of Ages # 26,
Abandoned Section, E.L. Smith Quarry,
Barre, Vermont,
1991

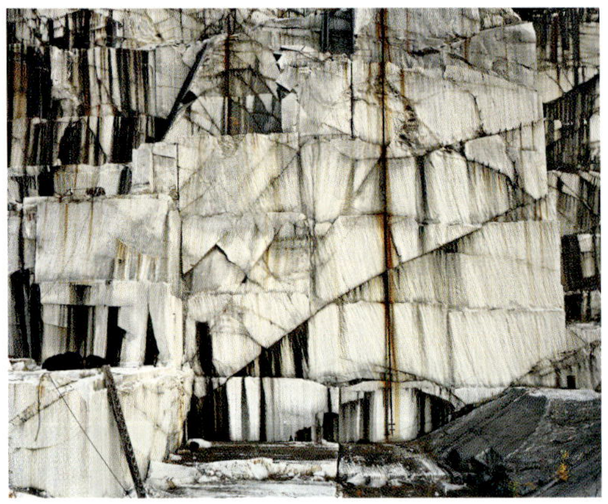

31

Rock of Ages # 10,
Abandoned Granite Quarry,
Barre, Vermont,
1991

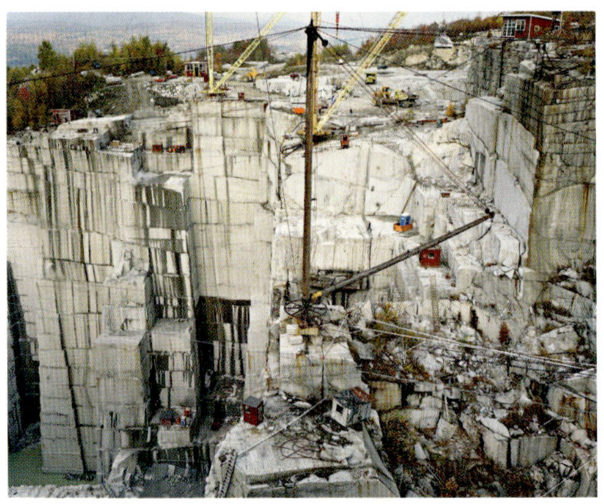

33

Rock of Ages # 19,
Granite Section, Rock of Ages Quarry,
Vermont,
1991

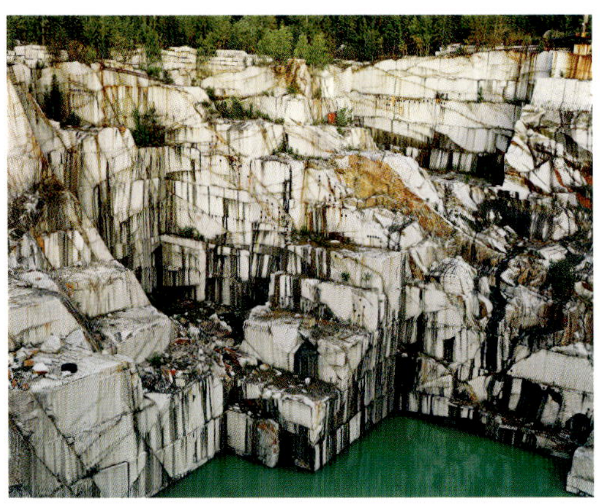

35

Rock of Ages # 23,
Abandoned Section, Adam-Pirie Quarry,
Barre, Vermont,
1991

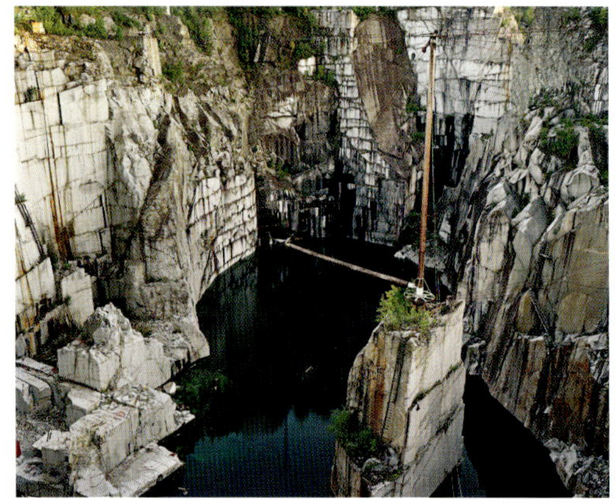

37

Rock of Ages # 22,
Abandoned Section, Wells-Lamson Quarry,
Barre, Vermont,
1991

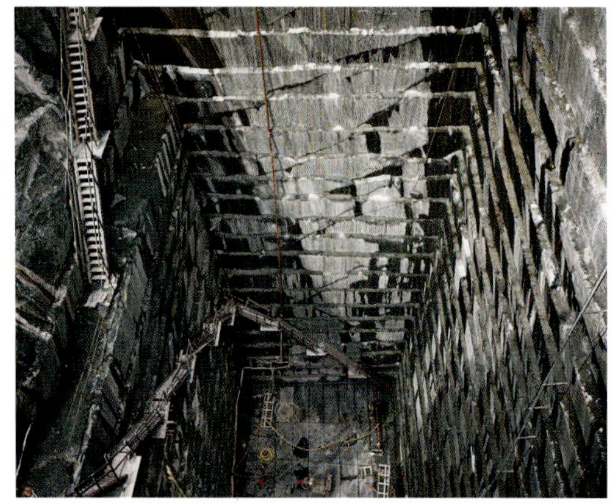

39

Vermont Marble Company # 52,
Vermont Marble Company,
Rochester, Vermont,
1991

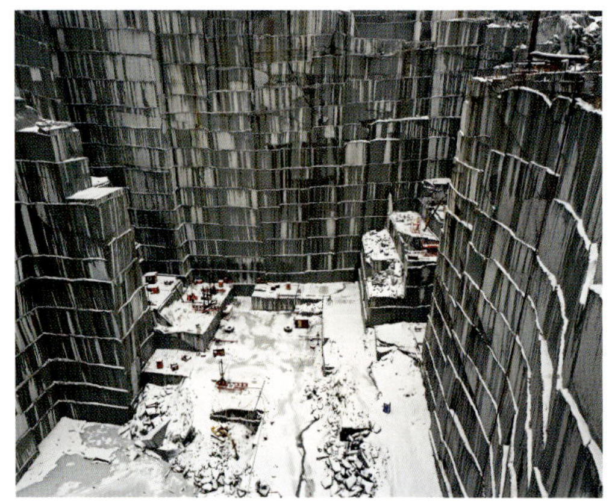

41

Rock of Ages # 15,
Active Section, E.L. Smith Quarry,
Barre, Vermont,
1992

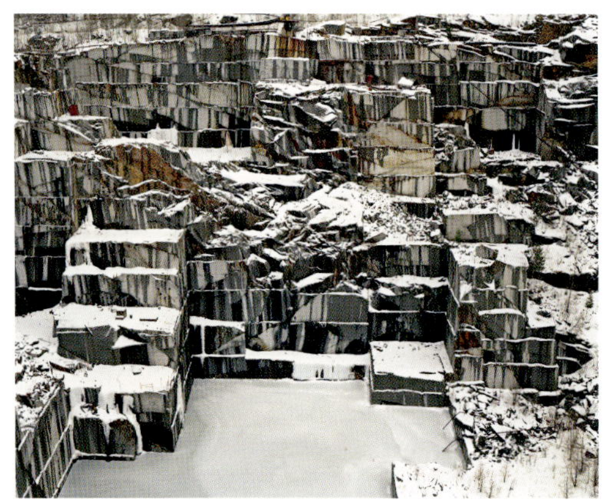

43

Rock of Ages # 14,
Active Section, E.L. Smith Quarry,
Barre, Vermont,
1992

ROCK OF AGES

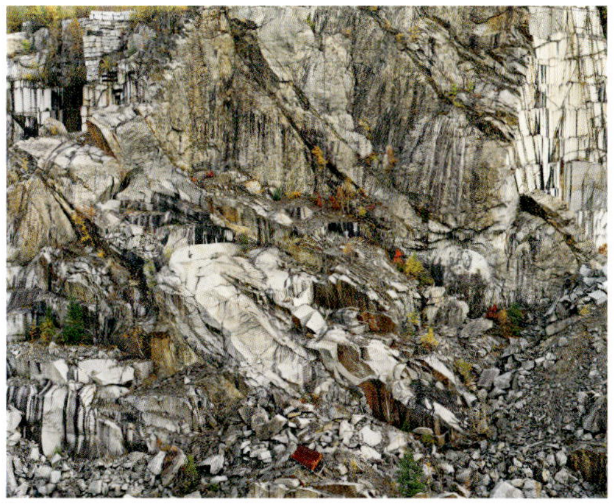

45

Rock of Ages # 6,
Abandoned Section, Rock of Ages Quarry,
Barre, Vermont,
1991

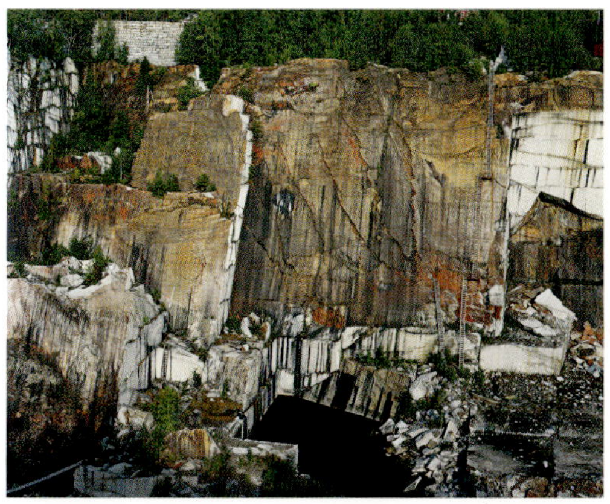

47

Rock of Ages # 45,
Abandoned Section, Rock of Ages Quarry,
Vermont,
1991

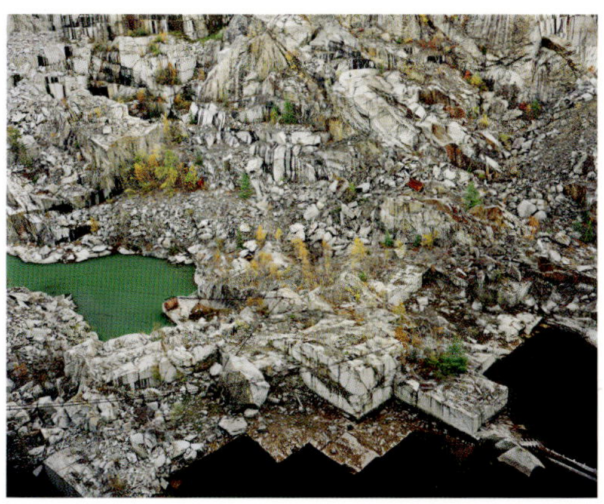

49

Rock of Ages # 24,
Abandoned Section, Rock of Ages Quarry,
Barre, Vermont,
1991

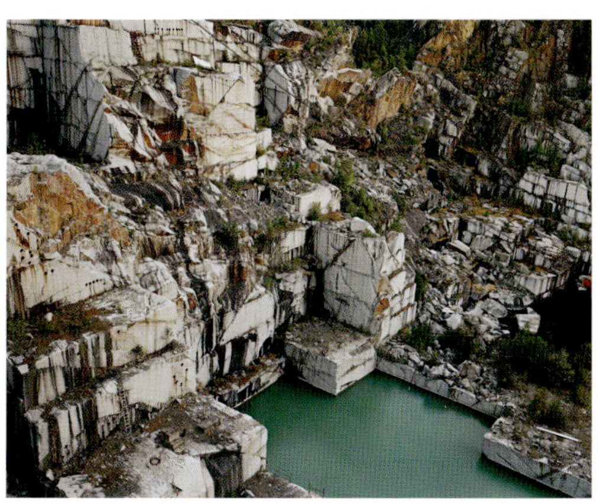

51

Rock of Ages # 59,
Abandoned Section, Adam-Pirie Quarry,
Barre, Vermont,
1991

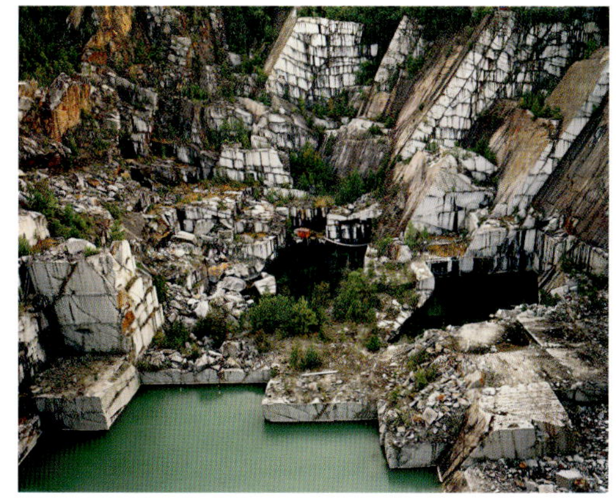

53

Rock of Ages # 25,
Abandoned Section, Adam-Pirie Quarry,
Barre, Vermont,
1991

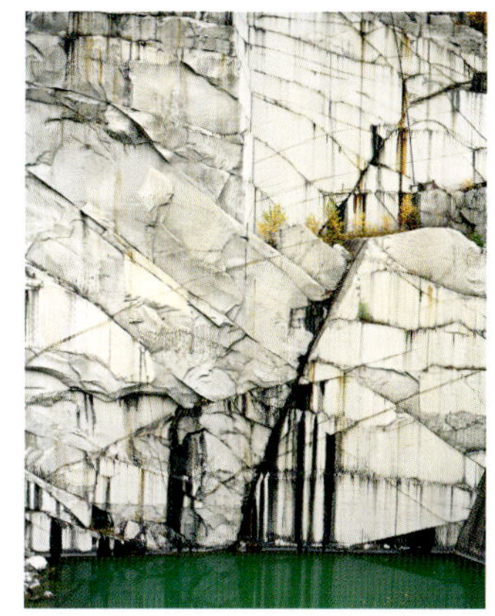

55

Rock of Ages # 4,
Abandoned Section, Adam-Pirie Quarry,
Barre, Vermont,
1991

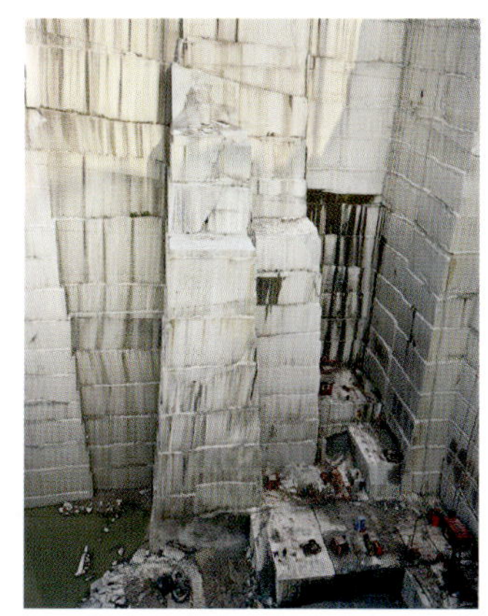

57

Rock of Ages # 39,
Active Section, E.L. Smith Quarry,
Barre, Vermont,
1991

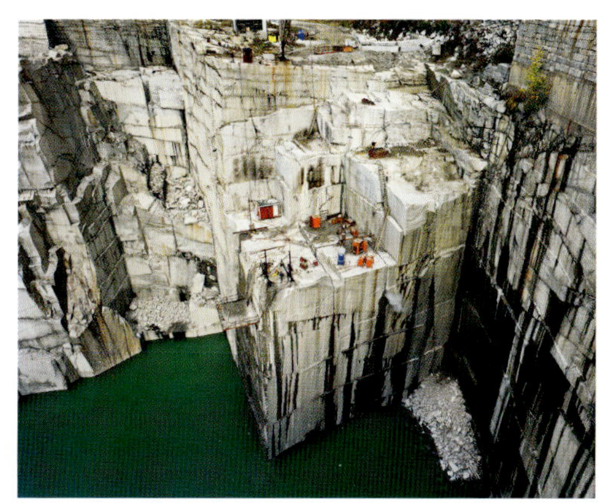

59

Rock of Ages # 7,
Active Granite Section, Wells-Lamson Quarry,
Barre, Vermont,
1991

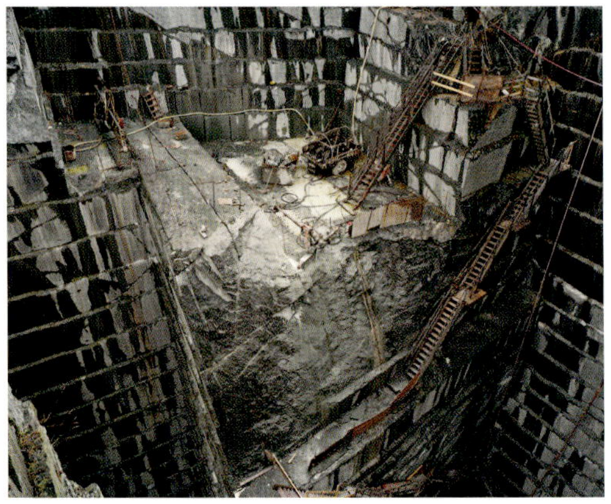
61

Vermont Marble Company # 16,
Active Marble Section,
Rochester, Vermont,
1991

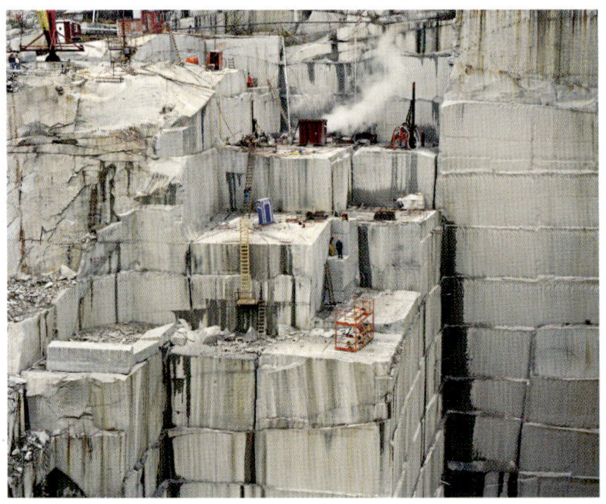
63

Rock of Ages # 1,
Active Section, E.L. Smith Quarry,
Barre, Vermont,
1991

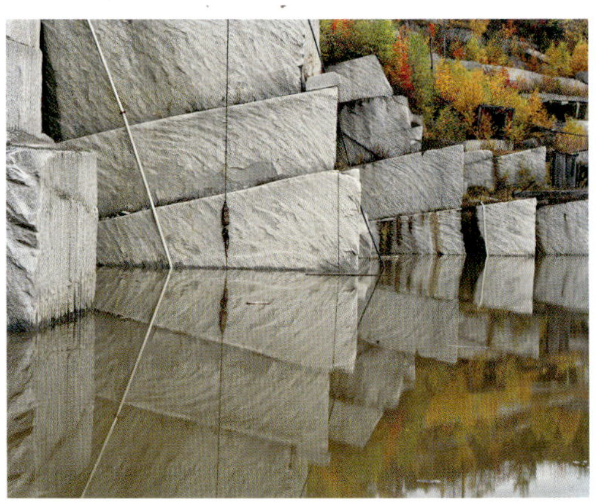
65

Rock of Ages # 2,
Granite Quarry,
Bebee, Quebec,
1991

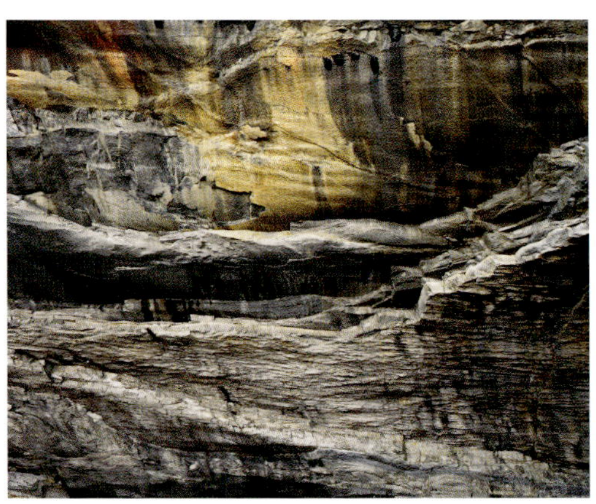
66

Abandoned Marble Quarry # 18,
Rultand, Vermont,
1991

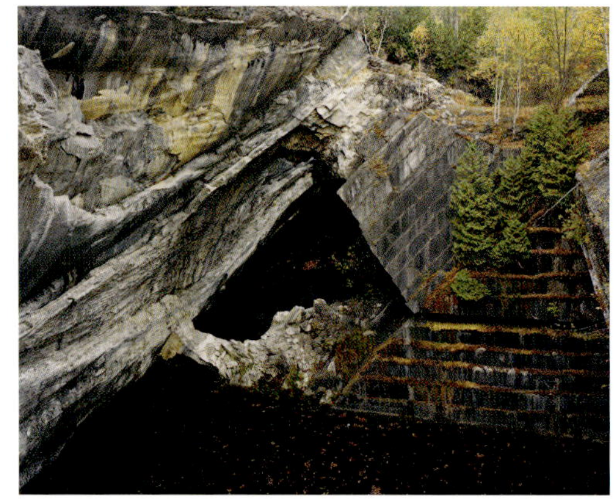
67

Marble Quarries # 001,
Rultand, Vermont,
1991

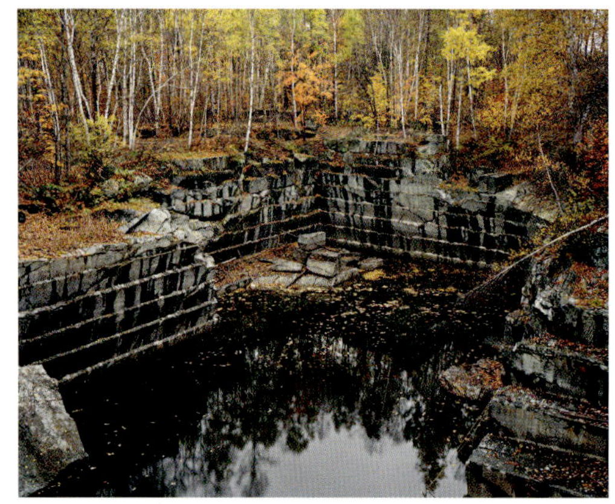
69

Vermont Marble Company # 5,
Abandoned Marble Quarry,
Rochester, Vermont,
1991

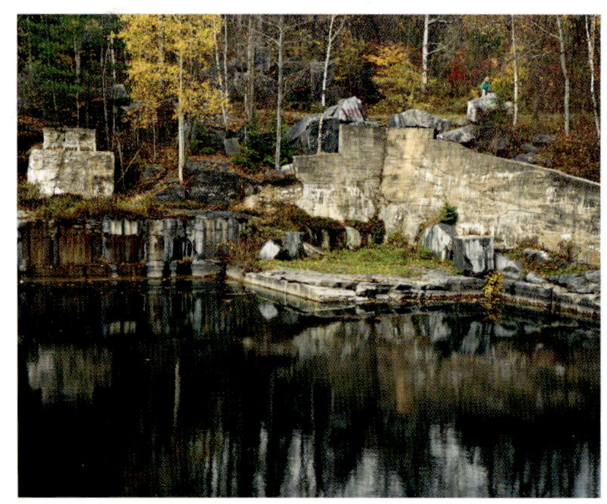
71

Rock of Ages # 3,
Abandoned Marble Quarry,
Near Rutland, Vermont,
1991

73

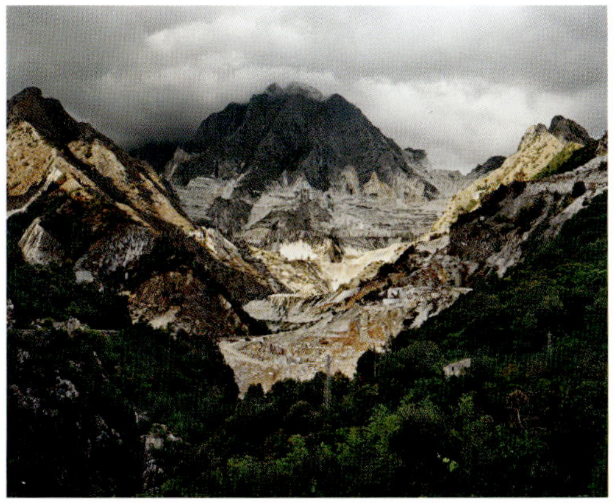

Carrara Marble Quarries # 20,
Carrara, Italy,
1993

75

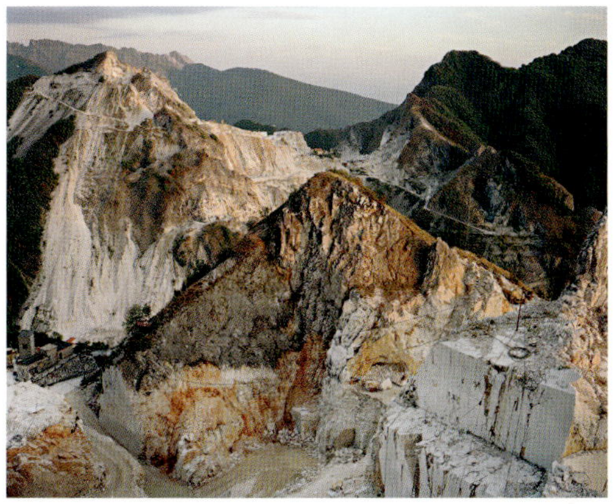

Carrara Marble Quarries # 22,
Carrara, Italy,
1993

77

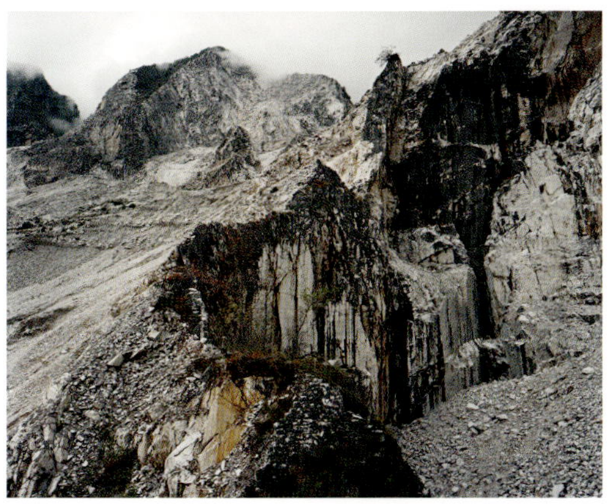

Carrara Marble Quarries # 23,
Carrara, Italy,
1993

79

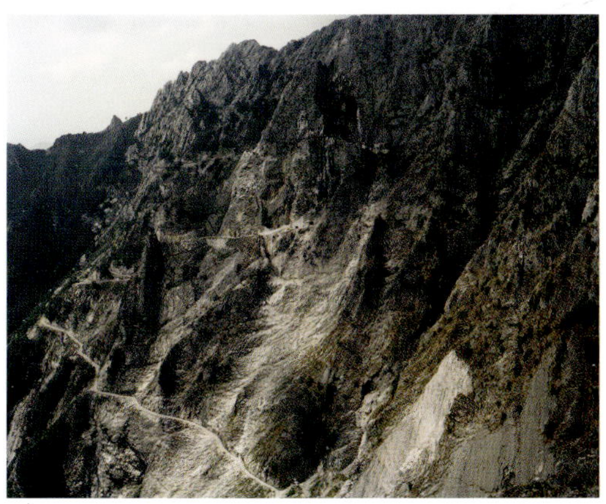

Carrara Marble Quarries # 15,
Carrara, Italy,
1993

81

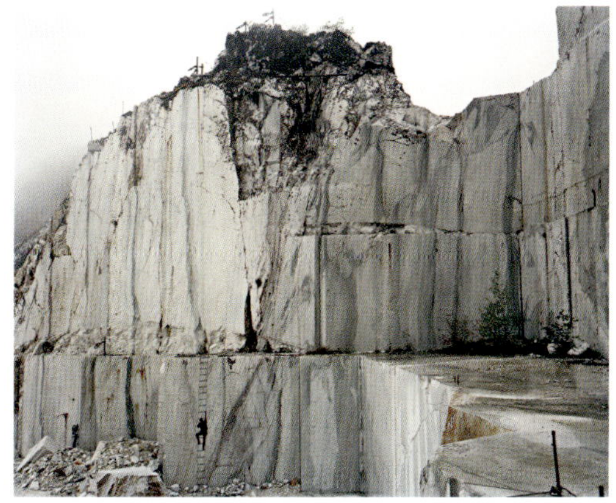

Carrara Marble Quarries # 11,
Carrara, Italy,
1993

83

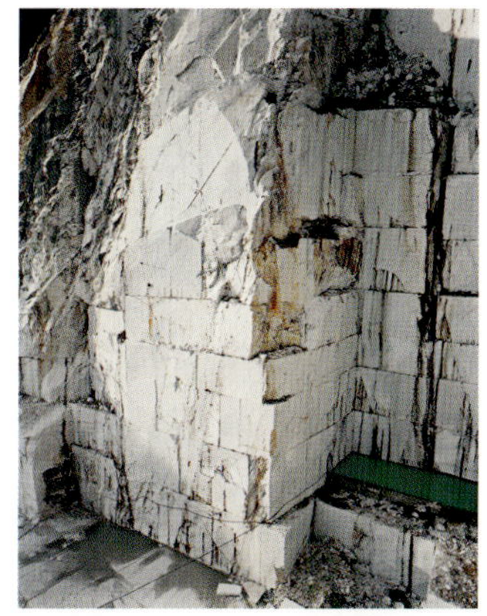

Carrara Marble Quarries # 12,
Carrara, Italy,
1993

85

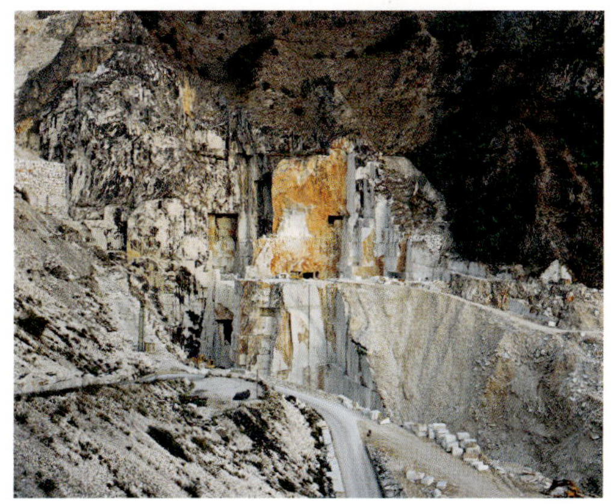

Carrara Marble Quarries # 31,
Carrara, Italy,
1993

87

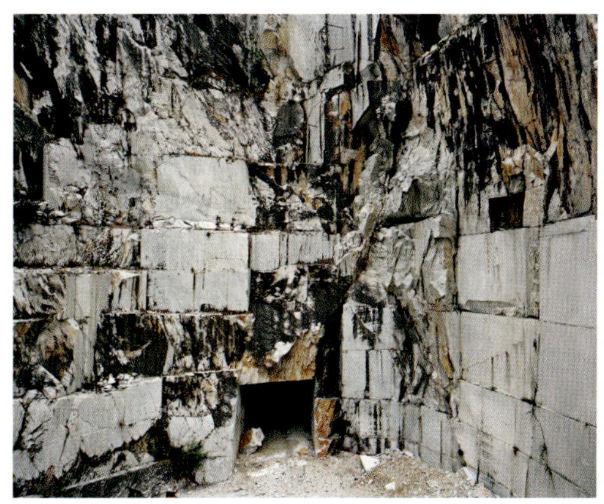

Carrara Marble Quarries # 14,
Carrara, Italy,
1993

CARRARA

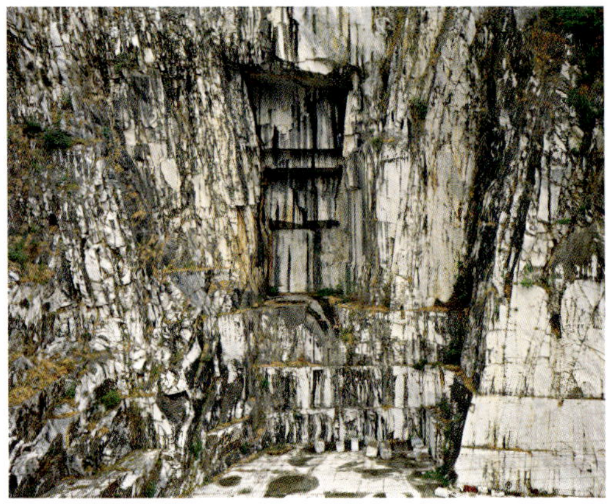
89

Carrara Marble Quarries # 3,
Carrara, Italy,
1993

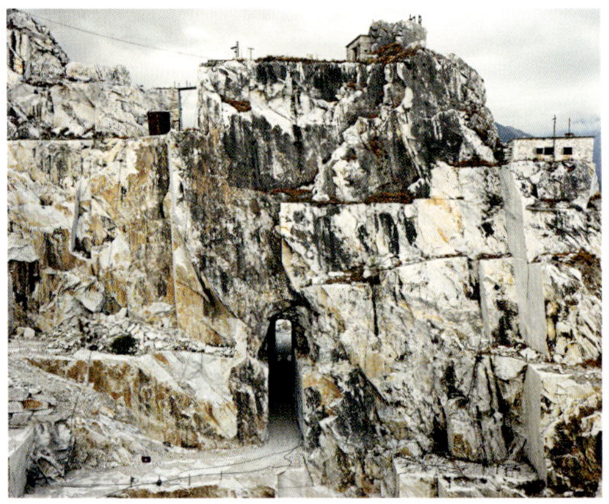
91

Carrara Marble Quarries # 30,
Carrara, Italy,
1993

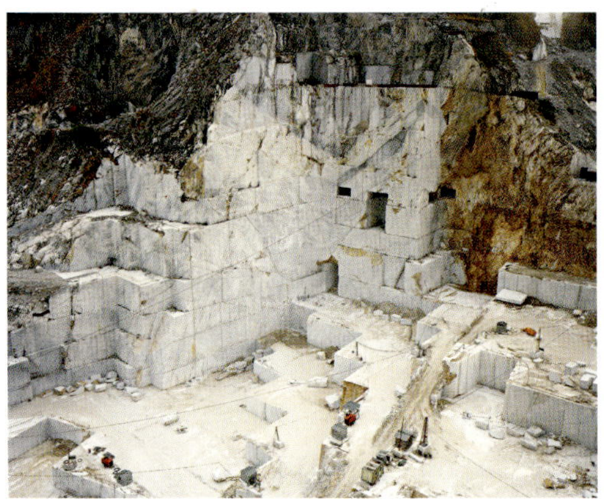
93

Carrara Marble Quarries # 4,
Carrara, Italy,
1993

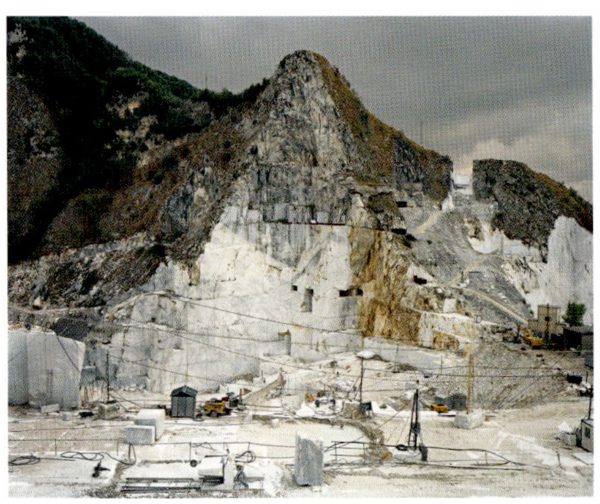
95

Carrara Marble Quarries # 13,
Carrara, Italy,
1993

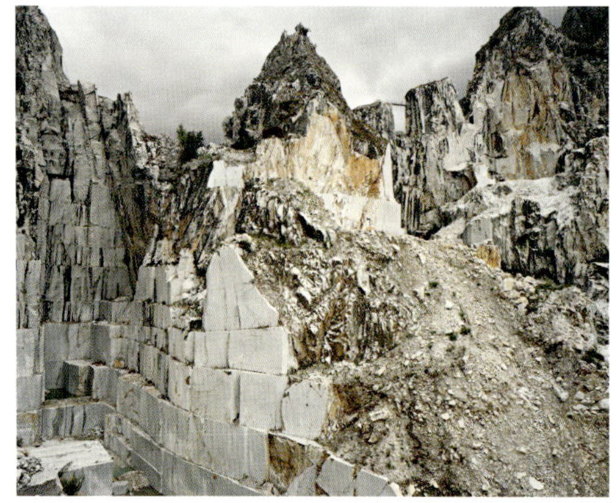
97

Carrara Marble Quarries # 32,
Carrara, Italy,
1993

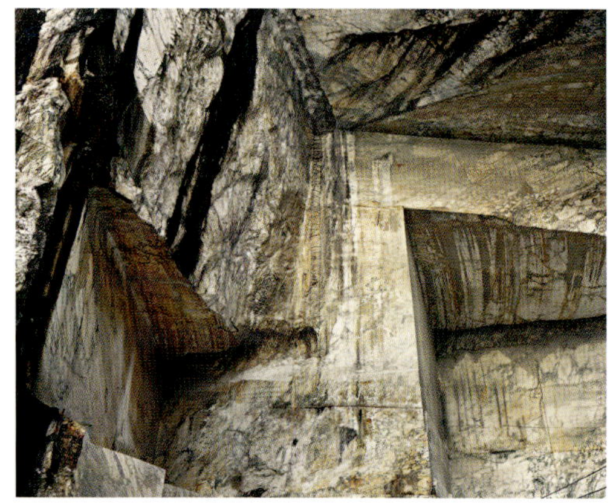
99

Carrara Marble Quarries # 21,
Carrara, Italy,
1993

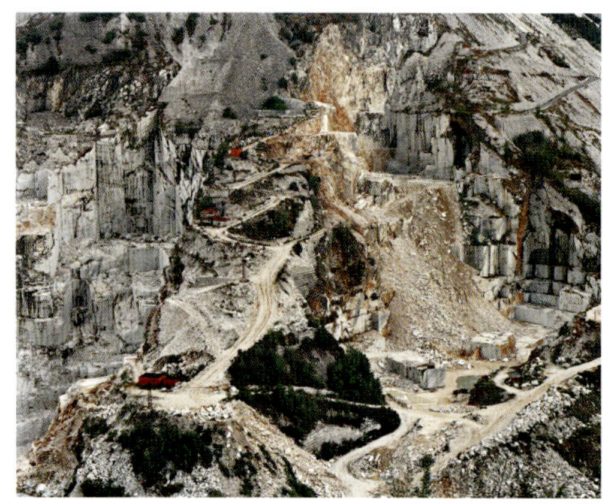
101

Carrara Marble Quarries # 2,
Carrara, Italy,
1993

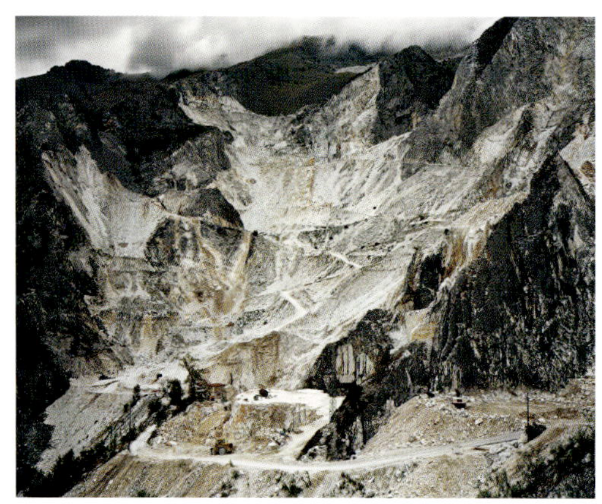
103

Carrara Marble Quarries # 28,
Carrara, Italy,
1993

CARRARA

104 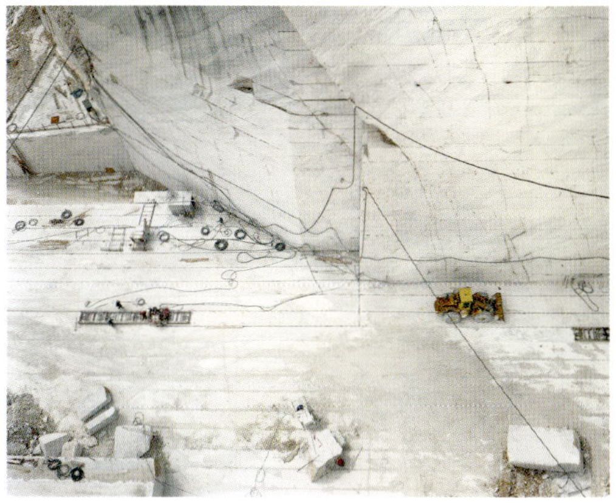

Carrara Marble Quarries # 24,
Carrara, Italy,
1993

105 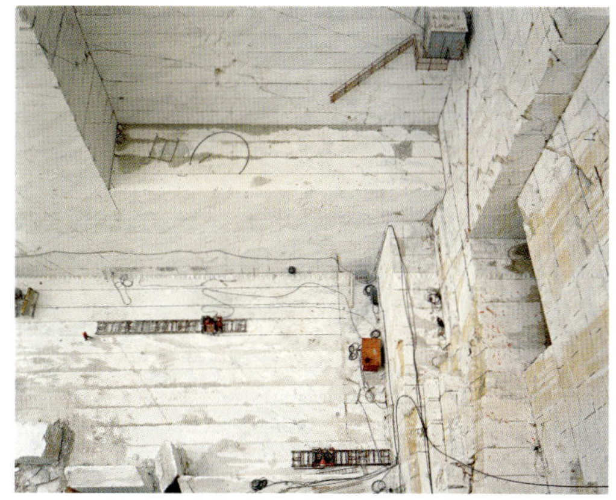

Carrara Marble Quarries # 25,
Carrara, Italy,
1993

MAKRANA

107 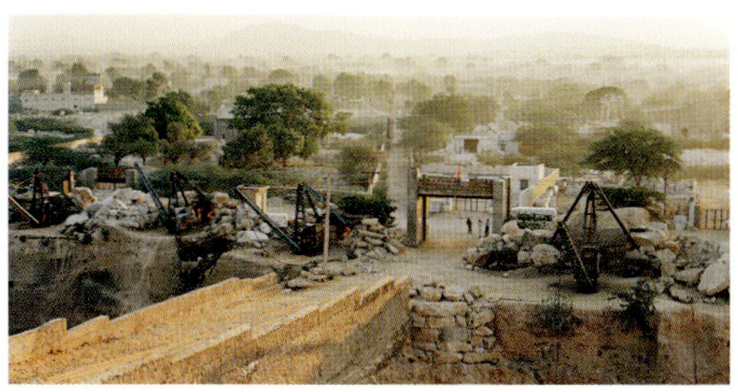

Makrana Marble Quarries # 18,
Rajasthan, India,
2000

111 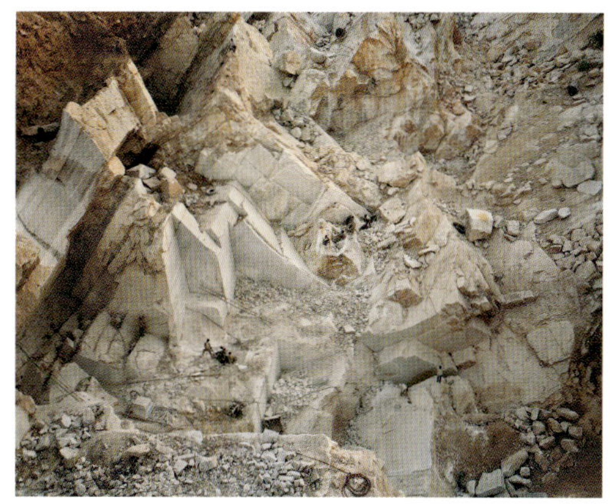

Makrana Marble Quarries # 11,
Rajasthan, India,
2000

109 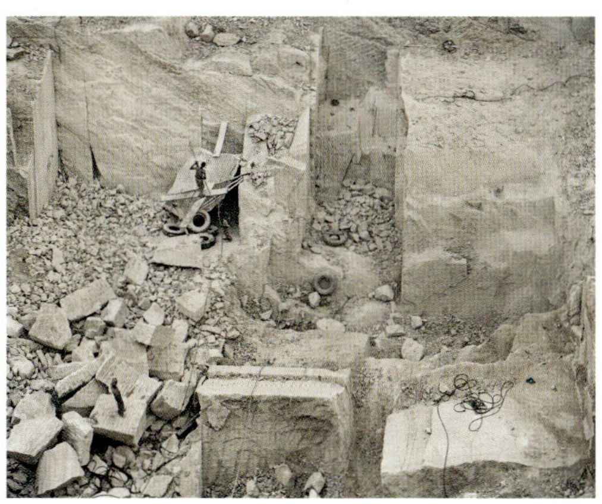

Makrana Marble Quarries # 13,
Rajasthan, India,
2000

113 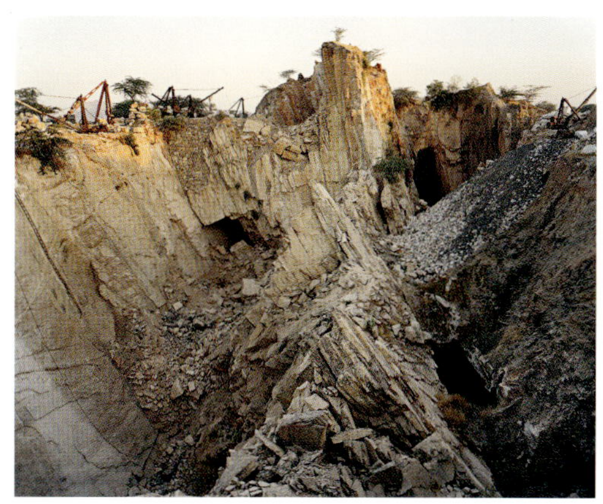

Makrana Marble Quarries # 8,
Rajasthan, India,
2000

MAKRANA

115 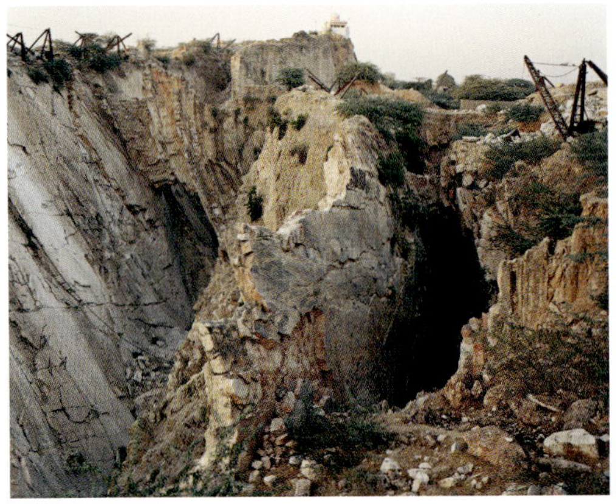

Makrana Marble Quarries # 7,
Rajasthan, India,
2000

117 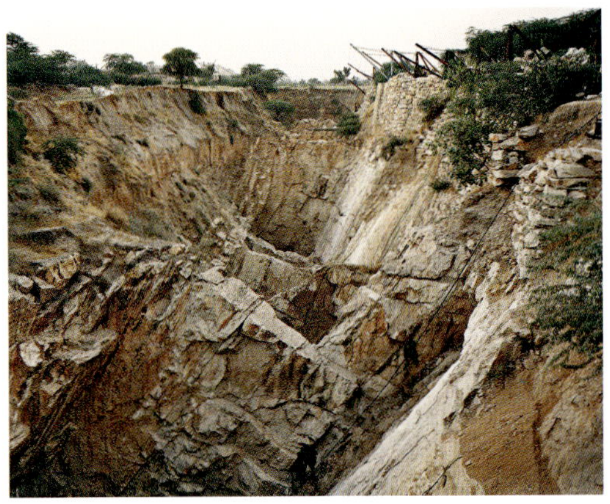

Makrana Marble Quarries # 9,
Rajasthan, India,
2000

119 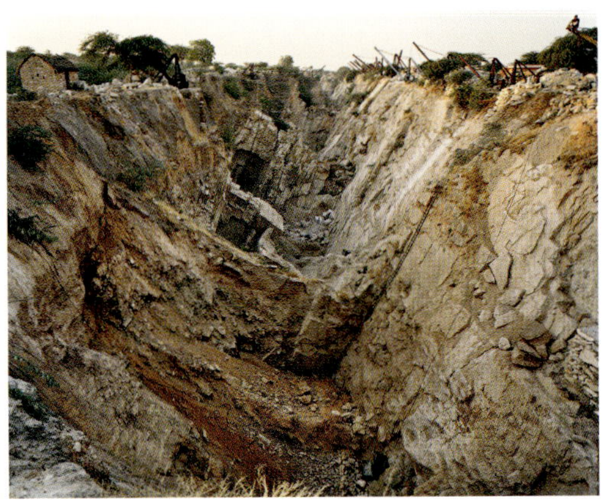

Makrana Marble Quarries # 5,
Rajasthan, India,
2000

121 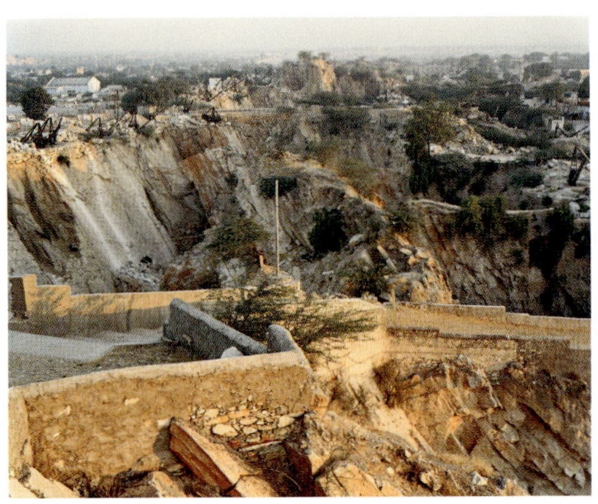

Makrana Marble Quarries # 10,
Rajasthan, India,
2000

123 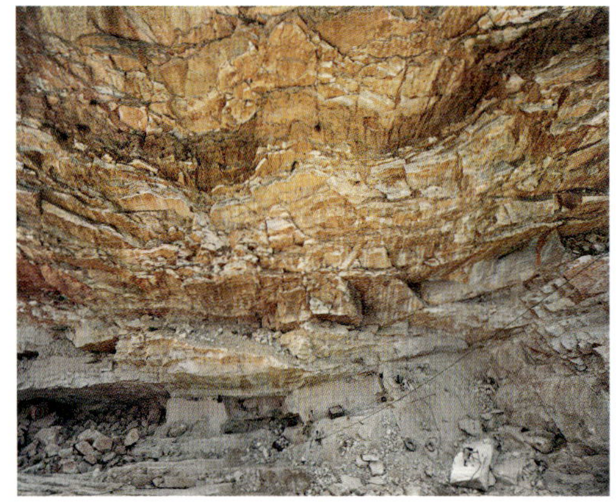

Makrana Marble Quarries # 6,
Rajasthan, India,
2000

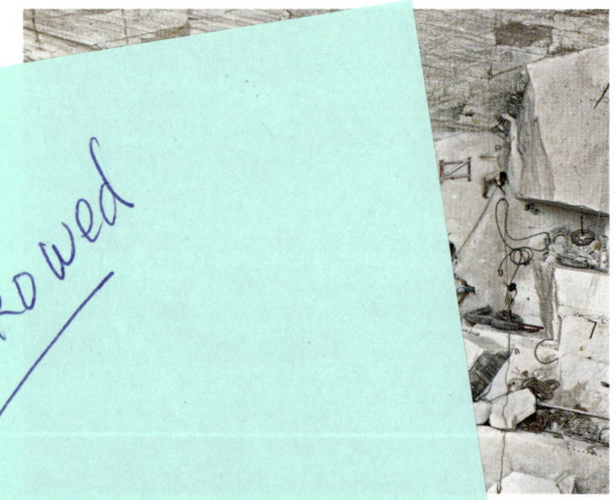

borrowed

Makrana Marble Quarries # 12,
Rajasthan, India,
2000

127 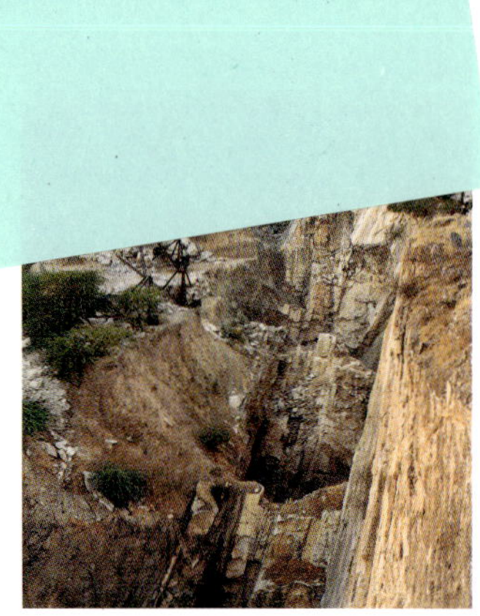

Makrana Marble Quarries # 1,
Rajasthan, India,
2000

129 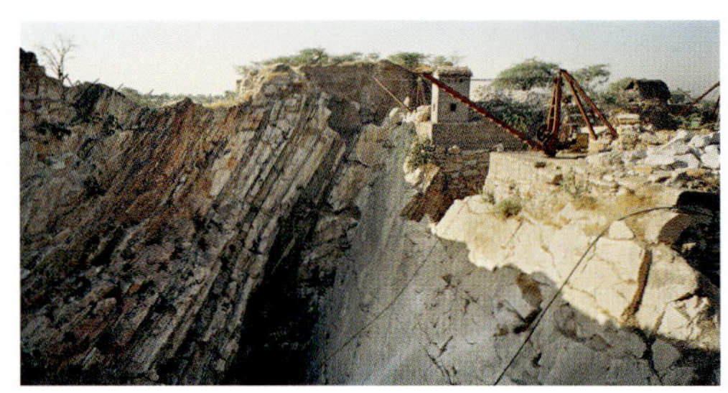

Makrana Marble Quarries # 17,
Rajasthan, India,
2000

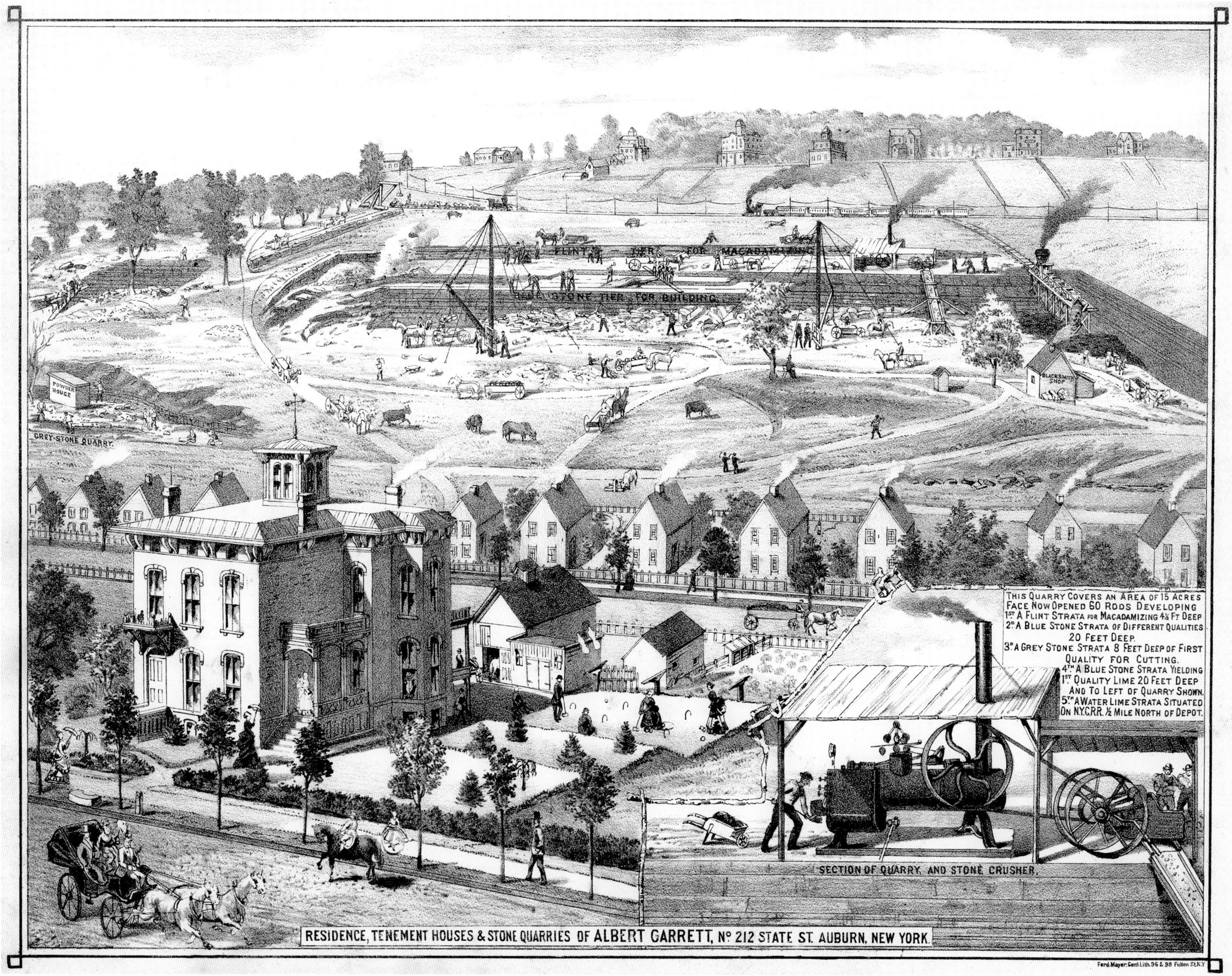

RESIDENCE, TENEMENT HOUSES & STONE QUARRIES OF ALBERT GARRETT, No 212 STATE ST. AUBURN, NEW YORK.

EDWARD BURTYNSKY · CURRICULUM VITAE

Born
St. Catharines, Ontario, 1955

Education
BAA Photographic Arts (Media Studies Program), Ryerson University, Toronto, Ontario.
Diploma, Graphic Arts, Niagara College, Welland, Ontario.

1985 to present - Photographic artist, entrepreneur, author, educator, lecturer.
Established Toronto Image Works, a darkroom rental facility, custom lab, digital imaging centre and new media computer training centre.

SELECTED SOLO EXHIBITIONS

2008
Edward Burtynsky: In the Pursuit of Progress, Winnipeg Art Gallery, Winnipeg, Manitoba
Edward Burtynsky: New Mines, Fotofreo Festival, Fremantle, Australia
Edward Burtynsky: The China Series (touring exhibition) Curated by David Brown - Southeastern Center for Contemporary Art (SECCA)
—Presentation House, Vancouver, British Columbia
—Tufts University Art Gallery, Medford, Massachusetts
—Samek Art Gallery, Bicknell University, Lewisburg, Pennsylvania

2007
Burtynsky: China, Center Koldo Mitxelena, San Sebastián, Spain
Burtynsky Photographs, (major survey) Gemeentemuseum Helmond, The Netherlands
Manufactured Landscapes, Canadian Cultural Centre, Paris, France
Art Gallery of Nova Scotia, Halifax & Yarmouth, Nova Scotia

2006
Edward Burtynsky: China, Palace of Art, Bratislava, Slovakia
Fabryka Krajobrazu/Manufactured Landscapes, Yours Gallery, Warsaw, Poland
Edward Burtynsky: The Edmonton Art Gallery Gift, Edmonton Art Gallery, Edmonton, Alberta
PhotoEspana, Centro Cultural de la Villa, Madrid, Spain

2005
Manufactured Landscapes, (includes 24 new China images), Brooklyn Museum of Art, Brooklyn, New York
Manufactured Landscapes, Museum of Photographic Arts, San Diego, California
Manufactured Landscapes, Iris & B. Gerald Cantor Center for Visual Art, Stanford University, Stanford, California
Entropia, Fundacion Bilbao Bizkaia Kutxa, Bilbao, Spain

2004
Manufactured Landscapes, Art Gallery of Ontario, Toronto
Manufactured Landscapes, Musée d'art Contemporain de Montréal

2003
Edward Burtynsky: Mid-Career Retrospective, The National Gallery ofCanada, Ottawa, Ontario
—Finnish Museum of Photography at Cable Factory, Helsinki, Finland
—The Art Gallery of Ontario, Toronto, Ontario
—The Brooklyn Museum of Art, Brooklyn, New York
—Musée d'Art Contemporain de Montreal
—The Museum of Photographic Arts, San Diego, California
—Iris & B. Gerald Cantor Center for Visual Arts, Stanford University

2000
Vues et points de Vue: L'architecture de Borromini dans les photographies d'Edward Burtynsky - Canadian Centre for Architecture, Montréal

1988
Breaking Ground - Touring exhibition produced by the Canadian Museum of Contemporary Photography:
—Art Gallery of Hamilton, Hamilton, Ontario - 1988
—Toronto Photographers Workshop, Toronto, Ontario - 1988
—Floating Gallery - Winnipeg, Manitoba - 1989
—Cedarbrae District Library, Scarborough, Ontario - 1989
—Whyte Museum of the Canadian Rockies, Banff, Alberta - 1990
—Bibliotèque municipale de Mont-Joli, Mont-Joli, Québec - 1990
—Alberta College of Art Gallery, Calgary, Alberta - 1991
—The Gallery/Stratford, Stratford, Ontario - 1991
—Galerie d'art de Matane, Matane, Québec - 1991
—Lacham Gallery, Stouffville, Ontario - 1992
—Cambridge Public Library and Art Gallery, Cambridge, Ontario - 1992
—Maltwood Art Museum and Gallery, Victoria, British Columbia - 1992
—Alberni Valley Museum, Port Alberni, British Columbia - 1992

1982
Landscapes & Greenhouses - Ryerson Photographic Arts Gallery, Toronto, Ontario

SELECTED GROUP EXHIBITIONS

2008
Ingenuity, Palais des Beaux Arts, Brussels, Belgium
Relics and Ruins, Instituto Oi Futuro, Rio de Janeiro, Brazil
China Works, Center Koldo Mitxelena, San Sebastian, Spain

2007
Imaging a Shattering Earth: Contemporary Photography and the Environmental Debate, CMCP, Ottawa
Utopian Mirage: Social Metaphors in Contemporary Photography and Film, Loeb Art Center, Poughkeepsie, NY
Moving Walls, Open Society Institute, New York
We are All Photographers Now!
Musée de l'Elysée, Lausanne, Switzerland
A History of Photography, Victoria and Albert Museum, London, U.K.
Imaging a Shattering Earth, Dalhousie Art Gallery, Halifax
Enfants De Cartier, Photographie Canadienne Contemporaine, Théâtre de la Photographie et de l'Image, Nice, France
Ingenuity, Gulbekian Foundation, Lisbon, Portugal

2006
C On Cities, Venice Bienalle, Pavillion Padiglione Italia, Giardini della Biennale, Venice, Italy
Imaging Asia in Documents, 1st Photo Biennale, Daegu , Korea
The photographic document: Aspects & Transformations, Photography Museum of Thessaloniki, Greece
Made in China, Museum of Contemporary Photography, Chicago, Illinois
Modern Times: Work, Machinery and Automation in the Arts of 1900, Palazzo Ducale, Genoa, Italy
Imaging a Shattering Earth, Museum of Contemporary Canadian Art (MOCCA), Contact Photo Festival, Toronto

2005
Imaging a Shattering Earth: Contemporary Photography and the Environmental Debate, Oakland University, Rochester, Michigan
Disaster Topographics, Toronto Photographer's Workshop, Toronto
Subjective Distance, McMaster Museum of Art, Hamilton, Ontario
Hot Mush and Cold North, The Ottawa Art Gallery, Ottawa, Ontario

2003
Industry and Entropy, Freedman Gallery, Albright College Center for the Arts, Reading, Pennsylvania (cat)

2002-2003

Altered Landscape: The Carol Franc Buck Collection,
Norsk Museum, Norway
Nevada Museum of Art, Reno

2002

New Acquisitions: New Work / New Directions3 / Contemporary Selections, Los Angeles County Museum of Art, California
Altered States: Landscape Transformation in the Wake of Progress, University Art Gallery, University of California, San Diego
Curator's Forum: Collecting Contemporary Art, Ackland Art Museum, University of North Carolina, Chapel Hill
The Eye of the Beholder: A History of Photography from the LaSalle Bank Photography Collection, Grand Rapids Art Museum, Michigan

2001

Detourism, The Renaissance Society, University of Chicago, Illinois
Site/Seeing, Art Gallery of Ontario, Toronto

1999

Virgin Land, Nassauischer KunstVerein, Weisbaden, Germany
L'Évocation, Marché Bonsecours, Montréal, Québec
The Sudbury Basin: Industrial Topographies,
The Art Gallery of Sudbury, Ontario

1998

On the Nature of Landscape, Mt. Holyoke College Art Museum,
South Hadley, Massachusetts

1997

Work, Workers, Works: Rearranging the Land,
Dalhousie University, Halifax, Nova Scotia
The Big Picture, Albright-Knox Gallery, Buffalo, New York
Land Used, Carleton College Art Gallery, Northfield, Minnesota

1996

Before the Land Behind the Camera,
Canadian Museum of Contemporary Photography, Ottawa

1994

Burtynsky/Maynard, Art Gallery of Ontario, Toronto

1993

Embodied Spaces, Le Mois de la Photo, Montreal, Quebec (cat)

1992

Beau: A Reflection on Beauty of Nature in Photography,
Canadian Museum of Contemporary Photography, Ottawa, (cat)

1991

Tainted Prospects: Photographers and the Compromised Environment,
Co-curated by John Pfahl and Sandra Olsen, Castellani Art Museum,
Lewiston, New York (cat)

1990

The Landscape: Eight Canadian Photographers,
McMichael Canadian Art Collection, Kleinburg, Ontario (cat)
Le Mois de la Photo, Curator: Katy McCormick, Concordia University,
Montréal, Québec

1989

Money Matters: A Critical Look at Bank Architecture,
Museum of Fine Arts, Houston, Texas (book)
Exhibition tour:
—Museum of Fine Arts, Houston, Texas - 1989
—National Building Museum, Washington, D.C. - 1990
—Canadian Centre for Architecture, Montréal, Québec - 1990
—Vancouver Museum, Vancouver, British Columbia - 1991
—National Gallery of Canada, Ottawa, Ontario - 1992
—Royal Ontario Museum, Toronto, Ontario - 1992
Structure and Landscape in Contemporary Canadian Photography,
Art Gallery of Northumberland, Cobourg, Ontario
What is Photography: 150 Years of Photography,
Prague, Czechoslovakia (cat)

1987

Focus '87, Art Gallery of Ontario, Toronto

1982/1983

Transitions: The Landscape,
Produced and toured by The National Film Board (now CMCP)
Exhibition tour:
—The Photo Gallery, Ottawa, Ontario - 1982
—Winnipeg Art Gallery, Winnipeg, Manitoba - 1982
—The Canadian Centre of Photography, Toronto, Ontario - 1983
—The Photography Gallery, La Jolla, S. California - 1983

AWARDS AND HONORS

Honorary Degree - Doctor of Laws, Queen's University, Kingston, Canada
Honorary Degree - Doctor of Fine Arts in Photography Study,
Ryerson University, Toronto
Honorary Degree - Doctor of Fine Arts, Montserrat College of Art, Boston
Officer of the Order of Canada
Flying Elephants Foundation Fellowship
TED Prize - (inaugural) Monterey, California
The Rencontres d'Arles Outreach Award
Roloff Beny Book Award, Toronto
Applied Arts Magazine - Photographic Book Award
Governor General of Canada Delegate – Circumpolar State Visit to Finland and Iceland
Royal Canadian Academy of Arts – Academician's Honours
National Magazine Awards Foundation – Photojournalism Silver Award

LECTURES

University of Toronto
National Geographic Society, Washington
ORACLE conference, Chicago
TED Conference, Monterey, California
Parsons Lecture, Parsons New School for Design, New York
Idea City, Toronto
Natural City Symposium, University of Toronto
International Center of Photography (ICP), New York
Finnish Museum of Photography at Cable Factory, Helsinki, Finland
International Center of Photography, New York
Library of Congress, Washington, D.C
Canadian Centre for Architecture, Montréal, Québec
National Gallery of Canada, Ottawa
Kodak Lecture Series, Ryerson Polytechnical University, Toronto, Ontario

SELECTED PUBLIC COLLECTIONS

Albright-Knox Art Gallery, Buffalo, New York
Art Gallery of Ontario, Toronto
Bibliothèque National, Paris, France
Brooklyn Museum of Art, Brooklyn, NY
Canada Council Art Bank, Ottawa
Canadian Centre for Architecture, Montréal, Québec
Canadian Museum of Contemporary Photography, Ottawa
George Eastman House, Rochester, NY
Gemeentemuseum Helmonds, The Netherlands
Glenbow Museum, Calgary, Alberta

Iris & B. Gerald Cantor Center for Visual Arts, Stanford University, Stanford, CA
Kitchener-Waterloo Art Gallery, Ontario
Library of Congress, Washington, D.C.
Los Angeles County Museum of Art
Musée d'art Contemporain de Montréal
Museum of Contemporary Canadian Art, Toronto
Museum of Fine Arts, Houston, TX
Museum of Photographic Arts, San Diego, CA
Museum of Modern Art, New York
National Archives of Canada, Ottawa
National Gallery of Art, Washington, D.C.
National Gallery of Canada, Ottawa
Ryerson University, School of Image Arts, Toronto
San Francisco Museum of Modern Art, CA
San Jose Museum of Art
Scottsdale Museum of Contempoary Art, Linhart Collection, Scottsdale, Arizona, U.S.A.
Solomon R. Guggenheim Museum, New York
Thessaloniki Museum of Art, Thessaloniki, Greece
Victoria and Albert Museum, London, England
Winnipeg Art Gallery, Winnipeg

SELECTED CORPORATE COLLECTIONS

Bank of Nova Scotia
Boston Red Sox
Ciba-Geigy Canada Ltd.
Canadian Imperial Bank of Commerce/Wood Gundy
Charles Schwab & Co.
Chase Bank
Dow Jones
Fidelity Investments
Foreign Affairs Canada
Goldman Sachs
J.P. Morgan Bank
Microsoft Corporation
The Parnassus Foundation
Power Corporation
Price Waterhouse
Royal Bank of Canada
Sunlife of Canada
Swiss Bank
Teleglobe Canada
Templeton Management
TransAlta, Corporation

SELECTED BIBLIOGRAPHY

2007

BURTYNSKY - QUARRIES, Published by Steidl, Germany, October, 192 Pages, Hardcover
TIME Magazine, July, New York
Mary-Kay Lombino, "Utopialand," Vassar College, New York
Jacobs, Karrie, "Searching For the Future," Metropolis, Architecture & Design, January, p.42
"La Bella Cicatriz / The Beautiful Scar," Ojo De Pez, The Fall of Nature, Issue #8, p.14-27
"Manufacturing Humanism: Steichen/Burtynsky," Prefix Photo #15, May, p.55-67
Callick, Rowan, "All the Coal in China," The American, May/June, p.24-32
"Photographie Canadienne Contemporaine," Théâtre De La Photographie Et De L'Image, Nice, May/Aug
"Age of Extraction," Walrus Magazine, July/August, p.32-34
"Histoire - Geographie Terminales," editions magnard, June, Paris
Vrij Nederland, Issue no. 17, Amsterdam, Holland
Daniel Canogar, "Ruins," Exit Magazine # 24, Nov 06/Jan, Madrid, Spain
"Worldchanging: A User's Guide for the 21st Century," January, New York, Climate Foresight

2006

Pozo Cruz, "The Seed of Destruction," La Fabrica, October, Madrid, Spain
Edward Burtynsky, "The Rice of China, Edward Burtynsky's view on China," Damn 6, Apr/May, p.22-29
Jules Petty, "Myth of Progress," Adbusters, #67 Sep/Oct, Vol 14, no. 5, Vancouver, p.70-71
Resurgence, May/June, Issue 236, Hartland, Bideford, UK.
National Geographic Society, September, Washington, "Work," hardbound book (2 images from the China series)
John K. Grande, "An Interview with Ed Burtynsky," Etc. Magazine, Issue #75 Autumn, Montreal
Allen and Mark, "Augmented Landscapes," Pamphlet Architecture 28, Princeton Architectural Press, New York
Viernes: El Periódico de Catalunya, April, Barcelona
Picnic 13, "Supervivencia y Bienestar / Survival and Well-being: China Portfolio," Nov/Dec, Mexico City
Sam Stephenson, "The Family of Destruction: Edward Burtynsky's Photographs," BardPolitik, p.35-38
Nancy Tousley, "Burtynsky in China," Calgary Herald, February 11, Calgary
Carol Diehl, "The Toxic Sublime," Art in America, February, New York, p.118
The Paris Review, no. 175, fall/winter, New York
Douglas Coupland, "Photography in Canada," Carte Blanche, Photography Vol. # 1, April 15, Toronto
Manfred Zollner, "Reich im Umbruch," fotoMAGAZIN, Issue #3, February 17, Hamburg
Stephanie Ringel, "Die Weltfabrik," Sie+Er Magazine, Nr. 1/06, January 2, Zurich
Andreas Weber, "Kuenstliche Landschaften: Die Schönheit des Hässlichen – Die Schönheit der Zerstörung," GEO Magazine, March 2006, Hamburg
C International Photo Magazine, Issue #2, June 2006 London, UK, Europe, North and South America, Asia
Mark Cheetham, "Art & Nature," The Walrus, April 2006, Toronto
Joanne Glaseby, "Fast Forward," Zag, Issue #2, March, London, New York
GEO Italia, Summer 2006, Milan

2005

Sansom, Anna, "The Beauty of Progress: Burtynsky's Widescreen Industrial Visions," November, Spoon, UK.
Meyers, William, "An Artist in Awe of His Surroundings," October 13, New York Sun, Arts & Letters, p.13
"Entropia," Fundacion Bilbao Bizkaia Kutxa, Bilbao, Spain, October (cat)
Baker, Kenneth, "As Nature Bleeds, Edward Burtynsky gets the Big Picture," July 6, San Francisco Chronicle

Knight, Christopher, "His Aim is True," May 25, Los Angeles Times
Dixon, Guy, "A Warning From the Wastelands," April 7, Globe and Mail, Review R1-R3
Goddard, Peter, "Terror in Progress," October 1, Toronto Star
Harris, Jeff, "The Reinvention of China," February 24, Maclean's, p.34-37
Roderick, Daffyd, "Made in Canada: Frame of Work" September 19, Arts & Culture, Time (Canadian Edition)
Milroy, Sarah, "Burtynsky's view of China Rising," October 15, Globe and Mail, Review, R5
"BURTYNSKY - CHINA," Published by Steidl, Germany, October, 148 Pages, 96 Colour Plates

2004
Dault, Gary Michael, "The Eleventh Hour in Photography," Summer, Canadian Art, Vol. 20 - No. 1, p.56
Fishman, Ted, "The Chinese Century," July 4, New York Times, p.24
Milroy, Sarah, "On the Eve of Destruction," May 31, Globe and Mail, R-6
Kuipers, Dean, "Guru of Wise Use," May, Playboy, p.51
Hannon, Gerald, "The Eyes of Edward Burtynsky," February, Toronto Life, p.81
Whyte, Murray, "Beautiful Mines," February, ARTnews, p.70
Whyte, Murray, "Adding up the price that Nature Pays," January, New York Times

2003
Richler, Noah, "This Overheating World: The Evidence of Man," Autumn, Granta, UK.
Dawson, Jessica, "Unnatural Wonders," August 7, Washington Post
Whyte, Murray, "Mapping the High Cost of Progress," June 28, Toronto Star, F-4
Darwent, Charles, "Unnatural Beauty," May 11, The Independent, p.23
Youngs, Christopher, "Industry and Entropy," March/April, Albright Center for the Arts (Cat.)
Fulford, Robert, "Industrial Light and Magic," March 11, National Post, Toronto

2002
Leffingwell, Edward, "Reviews: Edward Burtynsky at Charles Cowles," June, Art In America, p.129
Fulford, Robert, "Art Around Every Corner," May 9, The National Post, p. AL 3
Callahan, Sean, "Multiple Viewpoints," Smithsonian Magazine, April
"Real Art Ways, Acquiring Taste," Feb/June, p.85
Loke, Margarett, "Art in Review: Edward Burtynsky," The New York Times, February 22, p.E40

2001
Donegan, Rosemary, "Awesome Ambiguity: Edward Burtynsky's Industrial Landscape," Prefix Photo 3, p.9-23
Baker, Kenneth, "Edward Burtynsky," ARTnews, May
Richler, Noah, "Residual Landscapes," Saturday Night Magazine, May 19, p.25-29
Mays, John-Bentley, "A Melancholy Beauty," The National Post, April 25
Torosian, Michael, "Residual Landscapes: Studies of Industrial Transfiguration," Lumiere Press, Toronto
Mayer, Marc, "A Dystopian Sublime: Edward Burtynsky's Grand Tour," Montreal (cat)

1999
Murray, Joan, Canadian Art in the Twentieth Century, Dundurn Press
Hanna, Deirdre, "Ed Burtynsky ignites industrial passions," Now Magazine, Oct/Nov

1998
Long, Andrew, "Landscapes in Flux," Art & Antiques, December
Falk, Dan, "Mining modern industry for art gems," The Globe and Mail, May 5

1996
Rhodes, Richard, "Tailings," Canadian Art, vol. 13, number 3, Fall issue

1994
Burtynsky, Edward, Quarries, essay by Michael Mitchell (cat)

1991
Pfahl, John and Olsen, Sandra, "Tainted Prospects: Photographers and the Compromised Environment," Castellani Art Museum (cat)

1990
Wagg, Susan, Money Matters: A Critical Look at Bank Architecture, McGraw-Hill
Barbour, David, The Landscape, McMichael Art Collection (cat)
Harris, David, Money Matters: A Critical Look at Bank Architecture, (cat) Canadian Centre for Architecture

1988
Burtynsky, Edward, Breaking Ground, Canadian Museum of Contemporary Photography (cat)

1984
Langford, Martha, Contemporary Canadian Photography, Hurtig Publishers Ltd.

MEDIA

MANUFACTURED LANDSCAPES, a documentary by Jennifer Baichwal.

For complete information regarding the release of this award winning film and other media including television/radio interviews and news, visit: www.edwardburtynsky.com

ACKNOWLEDGEMENTS

As I look back on my career, a moment surfaces that brings me closer to an understanding of how fortunate I've been during my journey in the production of ideas and photographs. In my perfect world I wanted to create a full-service lab to produce my work and allow access to this technology for fellow artists/photographers. I started Toronto Image Works in 1985. After five years I hoped to hand over the day-to-day and resume my career as an artist. Life was not so perfect. By the late 1980's, a recession hit the Canadian economy and digital technology was lurking in the bushes ready to change photography at its core.

In 1989, I had just finished a commission to photograph bank architecture across North America for a project that was being assembled by the Houston Museum of Fine Arts. This project was the brainchild of Raphael Bernstein. In 1990, with the bank project coming to an end, Raphael came to Toronto Image Works to see how I was doing. I told him that I was afraid the business had got the better of me. He asked what I would photograph if I were not trapped in the routine of business. Without hesitation I blurted out, "Stone quarries… where stones are removed by the block. I want to find the structured and negative relief in the landscape where materials are harvested to build our towering cities." He thought it was a good idea, then said something very meaningful for which I will always be grateful. He said the work I did was important, and I would be doing the world a disservice should I stop. He offered to purchase ten of those yet unmade images. In the days that followed, Raphael sent magazines and articles about stone quarries and their locations. I responded to his generous offer, took a week here and there away from the lab and began making photographs again. The rest as they say, is history.

I continue to feel extremely fortunate to have had just the right person at the right time to turn my head and send me along a new path of discovery. That nudge put me back on course to my essential self, and his prescient vision about where my work might go has come to pass. Thank you Raf and Jane. You've become like a second set of parents who look after me.

Throughout those tough years I had another great supporter and ally, my life partner Jeannie Baxter. She held down the fort and kept things running at Toronto Image Works while I traversed the world in search of the subjects I wanted to form on my ground glass. I also wish to thank my two daughters Meagan and Katya for their patience and understanding during my absences. I would be remiss if I did not thank my mother Mary for her ongoing support. Her strength of character is an inspiration. Thanks too to Krys, Ollie, and Michael, my sisters and brother who are always there for me.

Quarries Book Project
Marcus Schubert – Editor, Art Director

Design
Jim Panou
Marcus Schubert

Colour
Dan Ebert

Field Assistance
Italy – Silvio Repetto
China – Noah Weinzweig
Iberia – Daniel Everett, Rosalind Williams,
Eva Yuste Ocaña

Editorial
Michael Mitchell – Essayist
Charles Oberdorf – Associate Editor
Kate Pocock – Associate Editor
Marcus Schubert – Contributor / Image research

Studio
Marcus Schubert – Production / Exhibitions Manager
Karen Machtinger and Nadia Belerique – Arts Administrators
Isabelle Martinez – Applied Production / Inventory Manager
Rose Scheler – Retoucher
Joseph Hartman – Intern

Toronto Image Works
Francine Lemieux – Customer Service
Ian Dollar – Digital Lab Director
Dan Ebert – Digital Colour Technician
Stephen Brookbank – Analog Printer
Tony Zhou – Mounting Technician

Special Thanks
I wish to extend my gratitude to all the quarry companies that took the time to make it possible for me to access and photograph their sites. Without their trust and cooperation this body of work would not exist.

Thanks to Nicholas Metiver for his friendship, sound advice and valuable guidance throughout my career.

Finally, I want to especially thank Marcus Schubert, Michael Mitchell, Jim Panou and Dan Ebert for their contributions and dedication to the production and design of this book.

Galleries
Nicholas Metiver – Toronto
Charles Cowles – New York
Robert Koch – San Francisco
Paul Kuhn – Calgary
Flowers East – London, U.K.
Toni Tapies – Barcelona
Stefan Röpke – Cologne
Art45 – Montreal

First edition 2007

Production: Bernard Fischer, Katja Töpfer, Gerhard Steidl
Printing: Steidl, Göttingen

Steidl
Düstere Str. 4 / D–37073 Göttingen
Phone +49 551-49 60 60 / Fax +49 551-49 60 649
E-mail: mail@steidl.de
www.steidlville.com / www.steidl.de

ISBN 978-3-86521-456-0
Printed in Germany

EDWARD BURTYNSKY · QUARRIES has been made carbon neutral by Zerofootprint. The carbon emissions associated with its production, printing and transportation have been offset with ISO-certified projects.

To learn more, visit: **www.zerofootprint.net**

The paper used in the production of this book is forest stewardship council (FSC) certified.